澜湄职业教育培训中心暨柬埔寨鲁班工坊系列教材
A Series of Textbooks for Lancang-Mekong Vocational Education Training Center and Cambodia Luban Workshop

电工——初级

Electrotechnology—Elementary Level

主编：张佳

Chief editor：ZHANG Jia

副主编：张链　张新江　李云龙

Deputy editors：ZHANG Lian　ZHANG Xinjiang　LI Yunlong

西安电子科技大学出版社

Brief Introduction

Based on the national vocational qualification standards for electrical technology and combined with the advanced skills training equipment in China, this book is compiled in a modular and task-oriented manner in terms of the application of related electrical technology primary skills. It is designed to provide comprehensive training in basic electrical circuit installation, comprehensive training in electrical control circuit, comprehensive training in analog electronic technology, and comprehensive training in digital electronic technology. Each module is designed with multiple typical work tasks. At the end of each module, 2 to 4 working tasks and 2 to 5 extension exercises are provided for readers who have the ability to expand their training or supplement their knowledge.

This book can be used as a basic textbook for applied undergraduates, higher vocational or open university students majoring in electric automation, mechatronics, energy engineering and electrical building. It can also be used as a reference book or skills training material for related technical personnel. Compiled in line with international teaching materials, it is designed in English for domestic readers who are seeking for professional English as well as overseas readers for their skills training or self-study.

图书在版编目(CIP)数据

电工＝Electrotechnology—Elementary Level：初级：英文 / 张佳主编. — 西安：西安电子科技大学出版社，2021.7
ISBN 978-7-5606-6122-3

Ⅰ. ①电…　Ⅱ. ①张…　Ⅲ. ①电工技术—教材—英文　Ⅵ. ①TM

中国版本图书馆 CIP 数据核字(2021)第 148371 号

策划编辑　明政珠
责任编辑　武翠琴
出版发行　西安电子科技大学出版社(西安市太白南路 2 号)
电　　话　(029)88202421　88201467　　邮　　编　710071
网　　址　www.xduph.com　　电子邮箱　xdupfxb001@163.com
经　　销　新华书店
印刷单位　广东虎彩云印刷有限公司
版　　次　2021 年 7 月第 1 版　　2021 年 7 月第 1 次印刷
开　　本　787 毫米 × 1092 毫米　1/16　印　张　10.5
字　　数　240 千字
定　　价　32.00 元
ISBN 978-7-5606-6122-3 / TM
XDUP 6424001-1

General Foreword

Serving the Belt and Road Initiative of China, the Lancang-Mekong Vocational Education Training Center and Cambodia Luban Workshop is a joint project undertaken by Tianjin Sino-German University of Applied Sciences(TSGUAS) for the Ministry of Foreign Affairs, the Ministry of Education and the Tianjin Municipal People's Government. Based in Cambodia, the project is designed to serve five countries in the Lancang-Mekong area and radiate to other ten ASEAN countries. It integrates functions of vocational training, vocational education, scientific research, cultural inheritance and innovation&entrepreneurship, develops both academic and non-academic education, and operates as a market-oriented international vocational training center.

At the initial stage of the project, 18 training rooms including mechanical processing technology, electrical technology and communication technology were built in three training centers for mechatronics and communication technology majors, with a total construction area of 6,814 m^2 and more than 1,600 sets of equipment.

The project will implement a "three-phase" plan. Based on the specialty construction in the first phase, international tourism, logistics engineering, automobile maintenance, building electricity and other specialties will be set up in the second phase to carry out technical skills training for Chinese&Cambodian enterprises and Cambodian people. Meanwhile, higher vocational education, applied technology undergraduate education, joint postgraduate education and other academic educations will be carried out to explore systematic talents cultivation of "medium and high vocational education, undergraduate education, and postgraduate education for a master's and doctoral degree".

Since 2017, as many as 95 articles about the project have been published by mainstream media including People's Daily, Guangming Daily, China Education News, Xinhuanet, etc. from home and abroad. After over two months of field study and research, Tianjin Television produced two feature stories named "Khmer Training", each lasting 30 minutes. The two episodes were broadcast on May 6th and May 13th 2019 respectively, featuring "on and on sails the vocational education, overseas shines the Luban Workshop". They give a full coverage of how TSGUAS teachers brought advanced skills to local areas and how friendship flourished along the Belt and Road Initiative route—a great contribution to the BRI. On July 18th, 2019, the Royal Government of Cambodia conferred the Officer of the SAHAMETREI Medal to the Secretary of the Party Committee of TSGUAS, and the Knight of the SAHAMETREI Medal to the President and Vice President in charge of this project, with the signature of Prime Minister Hun Sen of Cambodia. On July 22, 2019, China Education Association for International Exchange awarded TSGUAS the medal of "Featured Cooperation Project of China-ASEAN Higher Vocational Colleges". In October 2019, the President of National Polytechnic Institute of Cambodia (NPIC) presented 11 teachers with certificates and medals for their outstanding contributions to the Ministry of Labor and Vocational Training of Cambodia. Tianjin Sino-German University of Applied Sciences together with National Polytechnic Institute of Cambodia (NPIC) and their partners with enterprises was approved as the Belt and Road Joint Laboratory (Research Center)—Tianjin Sino-German and Cambodia Intelligent Motion Device and Communication Technology Promotion Center in December, 2020.

The Center has become a training base in Langcang-Mekong areas for technical talents training, a talent support base for Chinese enterprises overseas, a demonstration base for international students, and a base for teachers training. The Center is a key educational project of the Ministry of Foreign Affairs to serve the Belt and Road Initiative with foreign participation and entity institutions involved locally. The project will serve the social-economic and cultural development of the countries along the Initiative, enhancing the well-being of mankind; it will also serve the production output capacity of Chinese enterprises to help national development as well as enhance the international development of vocational education and the quality of its connotation. The project is a bridge connecting vocational education of Tianjin with the world, which marks a new stage of the city's international exchange and cooperation from a lower-medium to a medium-higher level.

The team of the project has compiled a series of textbooks for training, involving six occupations (electrotechnics, lathe, milling, CNC operation, bench and 4G communication network) from elementary, intermediate to advanced level based on current human resources situation in Langcang-Mekong countries, China's teaching equipment, and Chinese vocational qualification standards. These 19 textbooks target competence development and orient students to work tasks, combining theory with practice, and learning with practicing so as to put knowledge and skills into real situations. The textbooks aim to provide skills standards for the six occupations and lay foundations for the upgrading of the technological level of Lancang-Mekong countries.

ZHANG Xinghui

Party Secretary of Tianjin Sino-German University of Applied Sciences

March, 2021

A Series of Textbooks for Lancang-Mekong Vocational Education Training Center and Cambodia Luban Workshop Editorial Committee

Preface

In order to better serve China's "the Belt and Road" Initiative and export high-quality technical skills&educational concepts to the countries along Lancang-Mekong areas, teachers of the center have written this textbook based on the vocational skills standards and advanced equipment.

It emphasizes theoretical knowledge from national vocational qualification standards, combines the most advanced training equipment in the field of domestic vocational education, and adopts the systematic educational theories of the work process. All the knowledge and skills are fully reflected in each training module and specific work tasks. The book has made scientific and reasonable innovations and breakthroughs in content and structure. Guided by practical work tasks and focused on ability cultivation, the book is compiled according to the principle of project-based teaching mode, and combines theory with practice, and learning and practicing according to the actual task requirements, thus applying knowledge and skills into practice.

The book contains four training modules: comprehensive training in basic electrical circuit installation, comprehensive training in electrical control circuit, comprehensive training in analog electronic technology, and comprehensive training in digital electronic technology. Each module consists of 2 to 4 specific tasks, starting with basic theoretical knowledge. The textbook combines theories with practical working cases, so readers with or without professional foundation will feel free to understand what is dealt with. At the same time, this book provides some extended training tasks for the readers who have extra learning hours.

The training equipments involved in work tasks in this book are all based on the real training equipments of Lancang-Mekong Vocational Education Training Center and Cambodia Luban Workshop. The training modules and working content designed in this book are extracted from real cases and can meet all the regulations for primary technical knowledge and related skills requested in the "Electrician" national professional standards. At the same time, the book is compiled in English with simple and concise language which is in line with international textbooks. It can not only meet the learning needs of domestic readers who are seeking for corresponding professional English, but also can be applied to training and self-study overseas.

Zhang Jia of Tianjin Sino-German University of Applied Sciences acted as the chief editor and compiled the manuscript; Zhang Lian, Zhang Xinjiang and Li Yunlong of Tianjin Sino-German University of Applied Sciences acted as the deputy editors. Niu Yunsheng (Tianjin Sino-German University of Applied Sciences) gave guidance on the style and content of this book. And the teachers from National Polytechnic Institute of Cambodia also gave lots of advices during the book complication. As a result, we would like to express our heartfelt gratitude.

It is available to scan the two-dimensional code below for corresponding contents of this book in Chinese.

Due to our limited ability, the correction of any inappropriateness contained is welcomed by readers and peers.

Editors

January, 2021

Contents

Module One Comprehensive Training in Basic Electrical Circuit Installation

【General Introduction】

This project is designed for the elementary level electrician profession practitioners with two typical tasks. The main training objectives are recognition, practice, and basic operation. The contents include the installation, wiring and operation of indoor power distribution and lighting control circuits, and the installation, wiring and operation of backup power supply circuits for automatic switching of grid over-current. Through the practical training of this project, the trainees can meet the basic requirements for being electricians.

【Basic Elements】

1. Knowledge objectives

(1) Master the installation and wiring methods according to the circuit schematic diagram.

(2) Master the usage of a multimeter.

(3) Understand the working principle of the basic circuit.

(4) Master the working principle of the lighting circuit and automatic switching standby power supply circuit.

2. Competence objectives

(1) Be able to use a multimeter correctly.

(2) Through the learning tasks of indoor power distribution and the lighting control circuit and power supply circuit of power grid over-current automatic switching equipment, trainees master the installation, wiring and operation skills of power distribution and lighting circuit.

3. Operation standards

(1) Establish a group and divide the work.The experiment is usually carried out in small groups; each group is composed of 2-3 people. The experiment includes wiring, load adjustment, meter reading, data recording and other work. The members of the experimental group shall have a clear and reasonable division of labor, take positions and work in coordination. In addition, they shall carry out continuous consultations to ensure the standard and coordination of experimental operation.

(2) Select unit and instrument. Before the connection of the experiment, the appropriate unit and instrument (including the range grade of the instrument) should be selected according to the purpose of the experiment, and their corresponding models and parameters should be recorded.

(3) Graphical wiring. The connection method follows the following principle: series connection first and parallel connection next, that is, from the beginning to the end first link series into a main loop, and then link the parallel parts one by one. For the convenience of searching, it is better to use the wires of the same color for each route. After the completion of wiring, another member must check and verify before closing the switch for the experiment.

(4) Reading and recording. When reading instruments, attention should be paid to the ratio, minimum scale value, unit and effective number. If any abnormalities are found, cut off the power supply immediately, find out the cause and eliminate troubleshooting. After the experiment is finished, the experimental data are analyzed to check whether there are any abnormal points. If there are, the relevant data of this point are checked once again and corrected if errors are found.

(5) Removing wires. After the experiment, it is often the time when accidents are most likely to happen due to the relaxed mood. Therefore, the power supply must be turned off before disconnection and then more time must be spent to let off the charge stored in the capacitor before disconnection can be carried out. The wires, instruments and experimental units should be well arranged afterwards.

4. Safety standards

In order to ensure personal safety and equipment safety during the experiment, in addition to the reliable protective grounding, leakage protection, overload protection and safety plug-in wire and other safety measures, the experimental members must also strictly abide by the following safety operation procedures.

(1) During the experiment, human body shall not be exposed to charged body.

(2) Wiring and disconnection must be carried out with the power cut off.

(3) When switching on, all members of the experimental group must be informed and agree.

(4) In case of abnormal conditions during the experiment, such as smelling peculiar smell (such as the burnt smell of varnish), overheated unit or excessive vibration (or excessive operation noise), burnt fuse, sparks in the unit, and over full range (or reverse deviation) of the electricity meter, the power supply shall be cut off immediately to find out the cause and eliminate the fault.

(5) When the DC motor is directly started, it must be ensured that the excitation power has been connected. If the excitation power is not connected and the armature power is directly connected, it is easy to cause a runaway accident. Secondly, the variable resistor should be connected in series (started) in the armature circuit, and the sliding point of the rheostat should be placed at the place with the largest resistance value, or otherwise the armature circuit current will be too large (approximate short circuit) and damage the motor.

(6) Before the unit is powered on, check whether the motor can rotate flexibly (that is, check whether the rotor is stuck). If the rotor cannot rotate flexibly, readjust it to make it rotate flexibly.

(7) The main power supply of the laboratory shall be controlled by the experimental instructor, and other personnel can only operate with the permission of the instructor, and shall not switch on by themselves, which may cause major accidents.

(8) When using the resistor, it should be noted that the resistor cannot be placed at zero (or the position with a small resistance value), which is likely to burn out the resistor. In the experiment, the high value resistance is adjusted first and then the low value resistance is adjusted.

5. Quality standards

Analysis, research and discussion are carried out based on the measured data, the phenomena observed and the problems found in the experiment, and then we come to scientific conclusions. Clear thinking, concise writing, neat charts and graphs, reasonable analysis, clear conclusion, standard writing and smooth writing are the basic requirement for an experimental report.

【Training Requirements】

1. Preparation of tools, instruments and equipment

The tools, instruments and equipment required in the training of this module are shown in Table 1-1.

Table 1-1 Preparation table for tools, instruments and equipment

No.	Name	Unit	Quantity
1	Yalong YYL-WXD series training platform	Set	1
2	hanging board 1	Piece	1
3	hanging board 2	Piece	1
4	screwdriver	Piece	1
5	electrician pliers	Piece	1
6	wire stripping pliers	Piece	1
7	nipper pliers	Piece	1
8	multimeter	Piece	1
9	connecting test line	Set	1
10	attachment	Set	1

As shown in Figure 1-1, the hanging board 1 of the unit is equipped with leakage protection switch, fuse, time relay, indicator light, button and terminal board, which can provide main power switch, instruction switch, short circuit protection of circuit and indication signal of circuit operation for the training circuit.

As shown in Figure 1-2, the hanging board 2 of the unit is equipped with watt hour meter, ballast, fluorescent tube, time relay, starter, current relay, button switch, indicator light and terminal board. It can complete the practical training of the fluorescent lamp circuit, electric energy measurement circuit and unit distribution board circuit.

The unit hanging board 1 and the unit hanging board 2 can be used to add over-current protection to the control circuits of various motors, and can complete the practical training and assessment of lighting circuit connection in the electrician's post, primary and intermediate assessment.

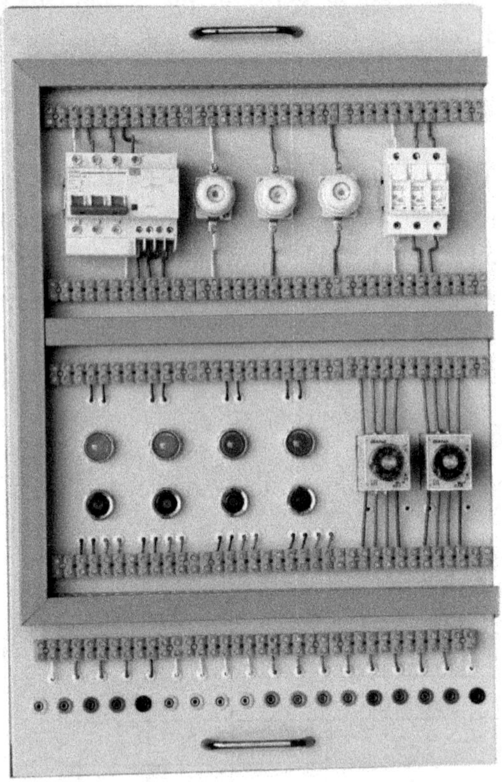

Fig.1-1　Hanging board 1

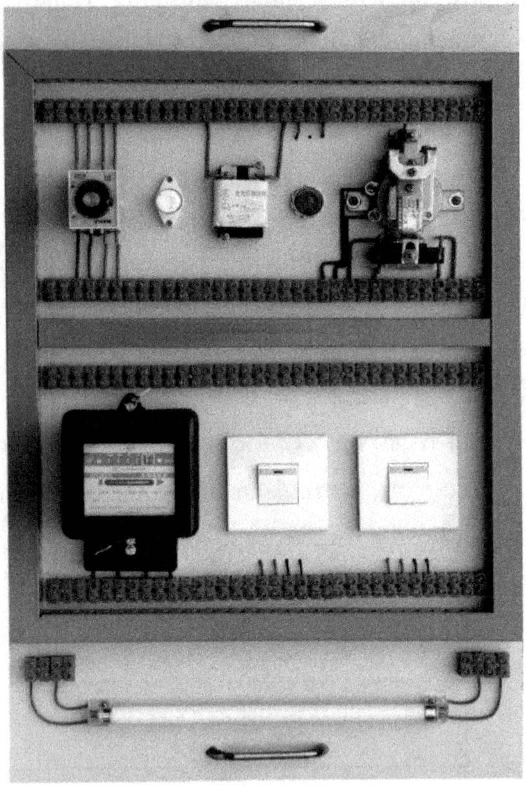

Fig.1-2　Hanging board 2

2. Knowledge assessment points

(1) Be able to define the working hours and process requirements according to the installation task list of lighting lines, and be able to divide labor.

(2) Be able to correctly identify and install the air switch, socket and other components, select the line color, and connect the wire with the wiring pile and the wiring cap.

3. Skill assessment points

(1) Be able to carry out slot board wiring construction with hand saw according to the drawings, technological requirements and safety regulations.

(2) Be able to test with a millimeter after construction according to the requirements of the construction specification; at the same time, be able to expand the skills of installation, wiring and operation of the lighting circuit and power grid over-current automatic switching standby power circuit.

【Work Tasks】

Task 1 Installation, Connection and Operation of Indoor Power Distribution and Lighting Control Circuit

【Relevant Knowledge】

1. Knowledge of electrical materials

1) Wire

Wires are also called conductors. Common conductors are divided into insulated conductors and bare conductors. The conductor core should have good conductivity, high mechanical strength, uniform texture, smooth surface, no cracks and good corrosion resistance. Conductor insulation wrapping is required to be of good insulation performance, flexible texture and considerable mechanical strength, and acid, oil, ozone corrosion resistance.

(1) Bare conductor.

The wires without their outside insulating layer become bare wires. Bare conductor is generally used for high-voltage wire erection in the field because it has no skin and is conducive to heat dissipation. In order to increase the tensile strength, the center of some aluminum strands is steel strand, which is called "steel cored aluminum wire".

Bare conductors are easy to cause accidents when they are used in densely populated areas due to their lack of insulation sheath. In cities where conditions permit, overhead high-voltage wires have been gradually converted to insulated wires or underground cables.

The material, shape and size of bare conductor are usually represented as follows: copper material is represented by letter "T"; aluminum material is represented by letter "L"; steel material is represented by letter "G"; hard material is represented by letter "Y"; soft material is represented by letter "R"; stranded wire is represented by letter "J"; section is represented by number; single wire diameter is represented by symbol "Φ". For example, LJ-35 represents aluminum twisted wires with a cross section of 35 mm^2. TΦ4 is a single stranded copper conductor with a diameter of 4 mm.

There are two kinds of common circular single wire: copper and aluminum. LY and LR are two kinds of aluminum single line, generally used for wires and cables, and copper single line is of two kinds, TY and TR.

Stranded wire is a kind of conductor made of multiple stranded wires. It aims to improve the performance. Bare stranded wire is a kind of stranded wire which is composed of several round wires. This kind of wire is soft and strong enough. Most of the overhead power lines and cable cores are stranded wires. The number of strands and the diameter of a single strand are expressed by putting the number of strands and the diameter together. For example, 7×2.11 means 7 strands of 2.11 mm diameter stranded wires. LJ type aluminum stranded wire, LGJ type steel cored aluminum stranded wire and TJ type hard copper stranded wire are commonly used in overhead lines. LJ and TJ types are mainly used for low-voltage and high-voltage overhead transmission, and LGJ type is mainly used for overhead transmission lines to improve tensile strength. The iron wire in bare stranded wire is often used in small power overhead lines, among which galvanized iron wire is widely used.

(2) Insulated conductor.

Wires with insulating cladding are called insulated wires. In terms of the core material, the insulated conductor can be divided into copper core and aluminum core; in terms of the number of core strands, it can be divided into single strand and multi-strand; in terms of the structure, it can be divided into single core, double core and multi-core; in terms of the insulation material, it can be divided into rubber insulated conductor and plastic insulated conductor.

The structure of rubber insulated conductor is to wrap a layer of rubber as the insulating layer outside the core, and then a layer of cotton yarn or glass fiber braid as the protective layer, which is mainly used for indoor wiring laying, and whose long-term working temperature shall not exceed 60℃. Rubber insulated conductor with AC voltage below 250 V can only be used for lighting circuit.

Plastic insulated wire uses PVC as the insulating coating, also known as plastic wire. Plastic wire has the characteristics of oil resistance, acid resistance, corrosion resistance, moisture-proof, mildew resistance, etc. It often acts as indoor lighting lines under 500 V, and can be directly laid on hollow slab or wall.

2) Cables

Cables are usually rope-like cables made up of several or groups of wires [at least two in each group]. Each group of wires is insulated from each other and often twisted around a center, with a highly insulated overburden covering the whole outside. The cables are mostly installed in the air or underground or underwater, and the insulated conductor composed of one or more mutually insulated conducting wires placed in a sealed sheath can be covered with a protective layer for telecommunication or power transmission. In 1832, Russian veteran officer buried the cable underground. The six wires were insulated by rubber and placed in glass tubes. This is the first underground cable in the world.

The advantages of cables over overhead lines are small insulation distance between lines, small space occupied, underground laying without occupying space above the ground, not affected by environmental pollution, high reliability of power transmission, and less interference to personal safety and the surrounding environment. But the cost is high, the construction and maintenance are more troublesome, and the manufacturing is more complicated. Therefore, cables are mostly used in densely populated areas, densely populated power grids and traffic congestion areas; large-span overhead lines can be avoided across rivers, and seabed. Cables can also be used in places where interference from overhead lines to communications needs to be avoided and where aesthetic considerations or exposure of objects need to be considered.According to the structure type, it can be divided into symmetrical communication cable and coaxial communication cable.

(1) Composition of the cable.

The cable mainly consists of the following four parts.

① Conductive wire core: made of high conductivity material (copper or aluminum). By the requirements of laying conditions for cable flexibility, each wire core may be made of a single wire or multiple wires.

② Insulation layer: The insulation material used for cable should have high insulation resistance, high breakdown electric field strength, low dielectric loss and low dielectric constant. The common insulating materials used in cables are oil-impregnated paper, polyvinyl chloride, polyethylene, cross-linked polyethylene, rubber and so on. Cables are often classified by insulation materials, such as oil-immersed paper insulated cables, polyvinyl chloride cables, cross-linked polyethylene cables, etc.

③ Sealing sheath: Protect insulating wire core from machinery, moisture, chemicals, light and other damage. Lead or aluminum extrusion sealing sheaths are generally used for moisture-prone insulation.

④ Protection cover: is used to protect the sealing sheath from mechanical damage. Generally, galvanized steel strip, steel wire or copper strip, copper wire, are used as armor to wrap around the sheath (called armored cable). The armor layer plays the role of electric field shielding and electromagnetic interference prevention. In order to avoid the corrosion of steel strips and steel wires by surrounding media, they are usually coated with asphalt or wrapped with impregnated jute or extruded with polyethylene and polyvinyl chloride.

(2) Classification of cables.

Cables can be divided into power cables, communication cables and control cables according to different uses.

① Communication cables. The types, specifications and main routes of communication cables are shown in Table 1-2.

Table1-2 Type, specification and main use of general cable

Product name	model	Section range / mm^2	Rated voltage (U_0/U)/V	Major uses
Light rubber sheathed cable	YQ	0.3~0.75 (2~3)	300/300	Applicable to household electrical installation, small electric mobile devices
	YQW			Same as above, oil and climate resistance (outdoor type)
Medium rubber sheathed cable	YZ	0.75~6 (2~5)	300/500	Various types of mobile electrical equipment
	YZW			Same as above, outdoor type
Heavy rubber sheathed cable	YC	2.5~120 (1~5)	450/750	All kinds of mobile electrical equipment, resistant to greater mechanical external force
	YCW			Same as above,outdoor type
Diving rubber sheath cable	YHS	0.75~6 (2~4)	450/750	Submersible motor

② Power cables. Specifications, models and main uses of power cables are shown in Table 1-3.

Table 1-3　Type, specification and main use of power cable

Name	Representative model	Specification	Main uses
Oil-immersed paper insulated cable package cladding type	ZQ，ZLQ， ZQ_{21}，ZQL_{21}	Voltage: 1-35 kV Section range: 2.5-240 mm^2	In the AC voltage transmission and distribution network for the transmission of electricity. Fixed in indoor, dry ditches and tunnels
	ZL，ZLL， ZL_{20}，ZLL_{20}	Voltage: 1-10 kV Section range: 10-500 mm^2	
Phase-Separated lead (Aluminum) cladding type	ZLLF，ZQF	Voltage: 20-35 kV	
Non-drip impregnated paper insulated cable package cladding type	ZQD_{31}，$ZLQD_{31}$ ZQD_{30}，$ZLQD_{39}$	Voltage: 1-10 kV Section range: 10-500 mm^2	Same as above, but usually used in high drop and vertical laying occasions
Phase-separated lead (Aluminum) cladding type	ZQDF，ZLLDF	Voltage: 20-35 kV	
Polyethylene insulated PVC sheathed cable	YV，YLV	Voltage: 6-220 kV Section range: 6-240 mm^2	Same as above, good corrosion resistance to the environment, laid indoors and in tunnels, not subject to external forces
Polyvinyl chloride insulated and sheathed cables	VV，VLV	Voltage: 1-10 kV Section range: 10-500 mm^2	
XLPE insulated PVC sheathed cable	YJV，YJLV	Voltage: 6-110 kV Section range: 16-500 mm^2 Multicore: 6-240 mm^2	Paper insulated cables with the same oil, but can be fixed laying for regular movement, without the limitation of laying position difference
Rubber insulated cable	XQ，XLQ， XLV，XV，XLF	Voltage: 0.5-6 kV Section range: 1-185 mm^2	Same with paper-impregnated insulated cables, but fixed laying for regular movement
Flame retardant XLPE insulated cable	YJT-FR (WD-YJT)	Voltage: 0.5-6 kV Section range: 1.5-240 mm^2	Flammable environment, commercial facilities, etc.

(3) Application of cable.

The application field of cable mainly includes power system, information transmission system, mechanical equipment and instrument system.

① The products of wires and cables used in the power system mainly include overhead bare wires, bus bars (buses), power cables (plastic cables, oil paper power cables which are basically replaced by plastic power cables, rubber sheathed cables, overhead insulated cables), branch cables (replacing some buses), electromagnetic wires and electrical equipment wires and cables for power equipment, etc.

② The wires and cables used in the information transmission system mainly include local telephone cables, television cables, electronic cables, radio frequency cables, optical fiber cables, data cables, electromagnetic cables, power communication cables or other composite cables.

③ In the mechanical equipment and instrument system, except overhead bare wires, almost all other products have found applications, mainly power cables, electromagnetic wires, data cables, instrument cables, etc.

2. Use of electrical tools

1) Pipe bender

The pipe bender is a special tool for bending circular tubes, as shown in Figure 1-3. The operation method for the pipe bender: place the PVC/metal pipe (part of the pipe needs) between the fixed wheel with guide groove and the fixed rod, then guide the pipe with guide groove of the movable rod, fasten the pipe with fixed rod, place the spring in the part of the pipe that needs bending, and the handle of the movable rod rotates smoothly clockwise. When operating, force should be applied slowly and steadily, bending with as large a radius as possible. Spring can keep the copper tube from being bent flat in a certain range, avoiding dead bending or cracking, as shown in Figure 1-4.

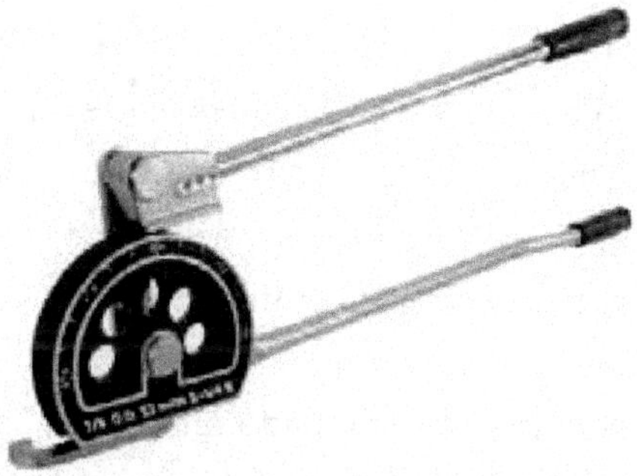

Fig. 1-3 Sample of pipe bender

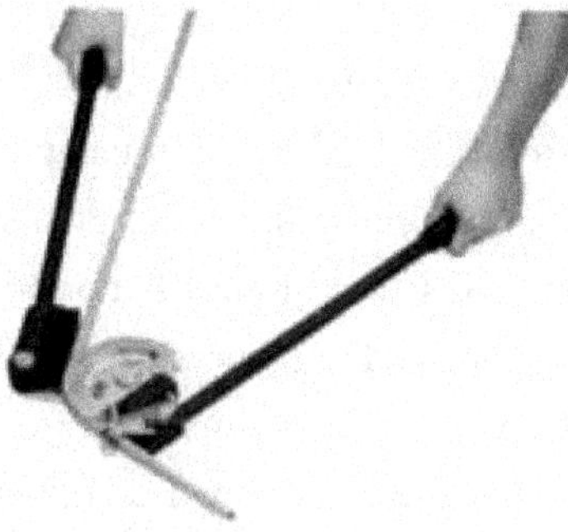

Fig. 1-4 Operational diagram of pipe bender

Precautions for using pipe bender:

(1) Dies must be selected according to the processing pipe diameter (or row size) and put them in place according to the serial number.

(2) No refueling shall be allowed between the pressurized pipe (row) and the die.

(3) Clamping machine parts, guide plate support mechanism should be timely reversed according to the direction of blended pipe (row).

(4) It is strictly forbidden for personnel to stay in front of the top die during operation and pressurization.

2) Herringbone ladder

Herringbone ladder, also known as folding ladder, as shown in Figure 1-5, is a kind of climbing tool used to work in the space above the plane. When it is used, the front and rear ladder poles and the ground form an isosceles triangle, which looks like a Chinese character "人", so it is called "herringbone ladder" in image.

The practical ladder falls into fixed ladder and movable ladder. Early ladders were usually made of wood. With the development of the metal industry, aluminum alloy materials are now widely used. Aluminum alloy ladders are light, beautiful and durable, and their cost is not high, so they are widely used.

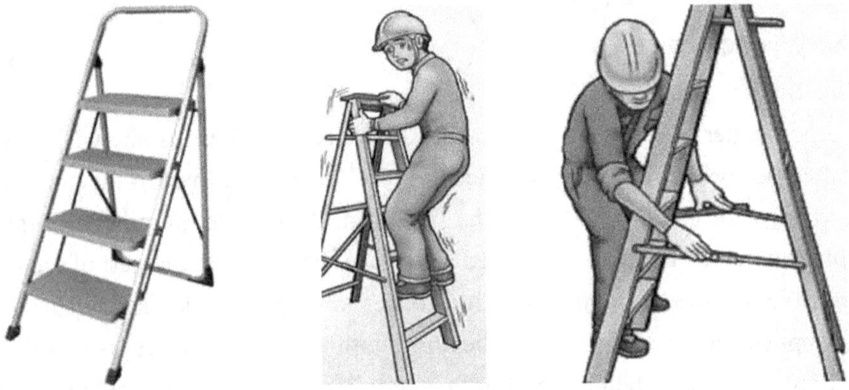

(a) Sample of herringbone ladder　(b) Moving ladder by foot　(c) Placing on a smooth surface

Fig. 1-5　Herringbone ladder

Precautions for using ladder:

(1) Use of the ladder: when using it, two people must be present, one person uses the ladder and the other person steadies the ladder; the bottom and foot of the ladder must be secured by reliable measures; the angle of the ladder should not be greater than 45 degrees; two people working on the ladder at the same time is prohibited; the operation on the top of the ladder is prohibited.

(2) In going up and down of the ladder, the person must face it, and no hand-held devices should be allowed; the bottom of the ladder foot should be strong, no pads should be used to raise the ladder, and no gear shortage should be used; the upper end of the ladder should be fixed and secured, and the lower end should be taken anti-skid measures; it is forbidden for two people to go up and down the ladder at the same time.

(3) If the ladder needs to be lengthened, reliable connecting measures must be ensured, and the joint should not exceed one.

(4) Operators should climb up and down within the prescribed passages.

(5) The ladder should not be used as a straight ladder.

(6) It is very dangerous to move the ladder "by foot" while standing on the miter ladder, as shown in Figure 1-5(b).

(7) Safety shoes should be worn and ensure that the soles are dry and non-slip.

(8) The user should face the ladder, with his body centered between the arms on both sides, and his three limbs should be able to touch the ladder at any time.

(9) Toolkits should be carried when working on the ladder to prevent falling objects.

(10) When using the ladder, the zipper with limited span should be fully opened and locked, and must be placed on a smooth surface, as shown in Figure 1-5(c).

(11) When the ladder is opened or closed, make sure the hand is away from the ladder winch and the ladder lock angle jacket.

(12) At least two steps away from the top of the ladder instead of stepping on the top of the ladder.

(13) The ladder should be placed in front of the work place, and the body should not be sideways, or otherwise it is easy to lose balance and fall.

3) Claw hammer with wooden handle

Claw hammer with wooden handle is a kind of hammer, as shown in Figure 1-6. Generally, one end of the claw hammer is round, while the other end is flat and bent downward with a V-opening, so as to lift the nail. Croissant hammers existed in ancient Rome, but the modern style of croissant hammers was designed by Americans.

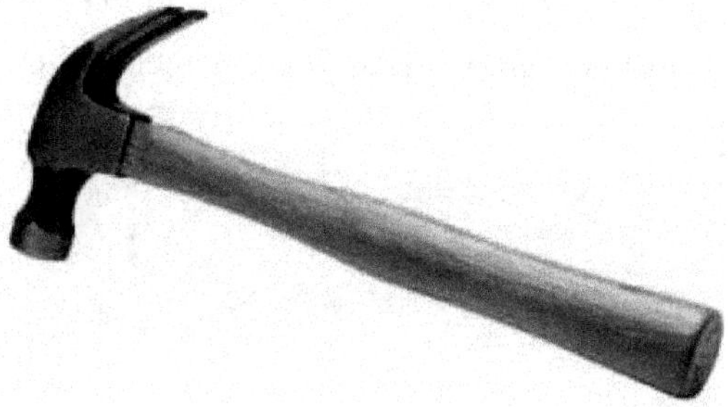

Fig. 1-6 Claw hammer with wooden handle

When using the hammer, we should note that the connection between the hammer head and the hammer handle must be firm. If it is loose, we should immediately tighten the hammer handle by wedge or replace it. The length of the hammer handle must be moderate. Experience shows that the appropriate length is when the length of the forearm holding the hammer is equal to the length of the hammer. Hand swing can be used when smaller striking force is needed. The method of arm swing should be adopted when strong striking force is needed, and the arc of the hammer head should be paid attention to when using arm swing and the handle of the hammer should not be contaminated by grease.

The following instructions should be noted when using the hammer:

(1) The connection between the hammer head and the handle must be firm. If the connection of the hammer head and the handle is loose, or if there is splitting or cracking on the hammer handle, the hammer must not be used. The length of the wedge should not be greater than 2/3 of the depth of the installation hole.

(2) In order to enable certain elasticity in hitting, the middle of the handle is slightly narrower than the end.

(3) When using a heavy hammer, attention must be paid to the front and back, left and right, and up and down. Standers are strictly prohibited within the scope of the heavy hammer movement, and no beating with a heavy hammer or a small hammer is allowed.

(4) The hammer head should not be quenched, and cracks and burrs should not be allowed.

The flying edge burrs should be repaired in time when they are found.

Sheep horn hammer can knock, hammer and pull nails, but it should not be used for hammering larger workpieces. When nailing, the hammer head should strike the nail cap flat so that the nail enters the wood vertically. When pulling out the nail, the hammer should be padded on the corner of the sheep to enhance the pulling force. The hammer should not be used as a pry. The hammer should be smooth and intact so as to prevent the nail flying out or the hammer slipping off and injuring others.

4) Hexagonal wrench

Hexagonal wrenches are also called Allen wrenches as shown in Figure 1-7. It differs from the inner hexagonal wrench and other common tools (such as a screwdriver and a cross screwdriver). It exerts force on the screw through the torque, which greatly saves the strength of the user. It can be said that, of the installation tools adopted in the modern furniture industry, the hexagonal wrench is not the most commonly used, but it is the best to use.

Fig. 1-7 Hexagonal wrench

Hexagonal wrench is used for assembling and disassembling large hexagonal bolts or nuts, which can be used by external wiring workers to load and unload steel frame structures such as iron towers.

5) Electrician's wire tong

As shown in Figure 1-8, electrician's wire tong is used to break and twist cylindrical metal parts and cut metal wires. Its side edges can also be used to cut fine metal wires. The steel wire clamp consists of a clamp head and a clamp handle. The clamp head includes a clamp, a tooth opening, a knife opening and a hydration opening.

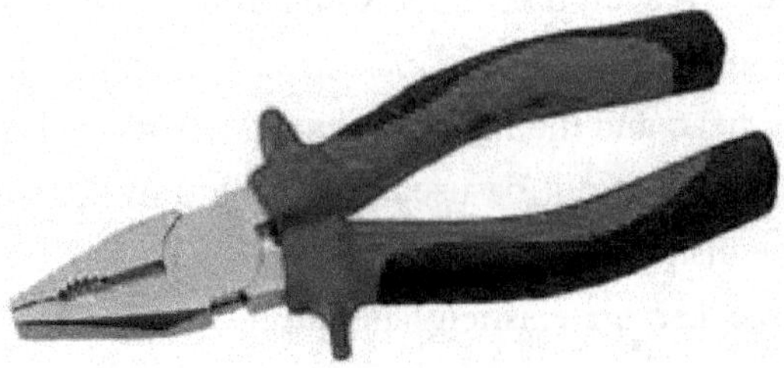

Fig.1-8 Electrician's wire tong

Electrical wire cutters are wire cutters certified by VDE with a voltage of 1000 V of insulation sleeve. It is used for clamping or breaking metal sheet and cutting wire. The following instructions should be noted when using it.

(1) Before using electric wire pliers, it is necessary to check whether the insulation of the insulating handle is in good condition. If the insulation is damaged, it is very dangerous to carry out live work, and an electric shock accident will occur.

(2) When shearing live wires with electric wire pliers, do not cut both the fire line and the zero line at the same time with the knife edge, so as to avoid short-circuit fault.

(3) Pay attention to the safe distance of the metal part of the clamp from the charged body when working on live.

6) Sharp-nose plier

As shown in Figure 1-9, sharp-nose plier is composed of a pointed head, a knife edge and a clamp handle. The material of which the electrical pointed clamp is made is $45^{\#}$ steel, which is classified as medium carbon steel. Its carbon content is 0.45%, being of suitable toughness and hardness. The clamp handle is covered by an insulating sleeve with a rated voltage of 500 V. It is a common clamp tool. The utility model is mainly used for shearing single and multi-strand wires of small wire diameters, bending rings for single wire joints, and stripping plastic insulation layer, etc. It can be operated in narrow workspace. Those without edge can only pinch workpiece, and those with edge can cut small parts.

The electric pointed nose pliers are one of the common tools in the assembly and repair of the internal line equipment. According to the length of the tong head, they can be divided into two kinds: short pliers (the pliers are 1 / 5 of the whole length of the pliers) and long pliers (the pliers are 2 / 5 of the whole length of the pliers). There are four specifications, namely 125 mm, 140 mm, 160 mm and 200 mm in length.

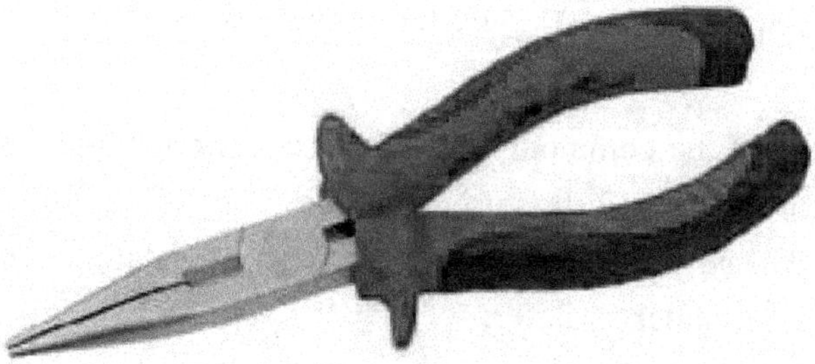

Fig. 1-9 Sample of sharp-nose plier

7) Diagonal plier

As shown in Figure 1-10, the diagonal plier is 1000 V inclined clamp certified by VDE. The pliers are mainly used for shearing conductors, and spare lead of components, and are often used to replace general scissors for shearing insulating bushing, nylon tie wire clamp, etc. The blade of the beveling pliers can be used to cut the rubber or plastic insulation layer of soft wires. The knife edge of the pliers can also be used to cut wires. When shearing No. 8 galvanized wire, cut it back and forth around the surface with a knife edge, and then cut the wire with a gentle pull. It can also be used to cut off hard wires such as wires and steel wires. There are 150 mm, 175 mm, 200 mm and 250 mm types commonly used by electricians. They can be purchased according to the type of work inside or outside. The jaws of pliers can also be used to tighten or loosen nuts. Inclined pliers are one of the hardware tools, also known as "inclined pliers". They find wide applications, and they are indispensable tools in daily life and work.

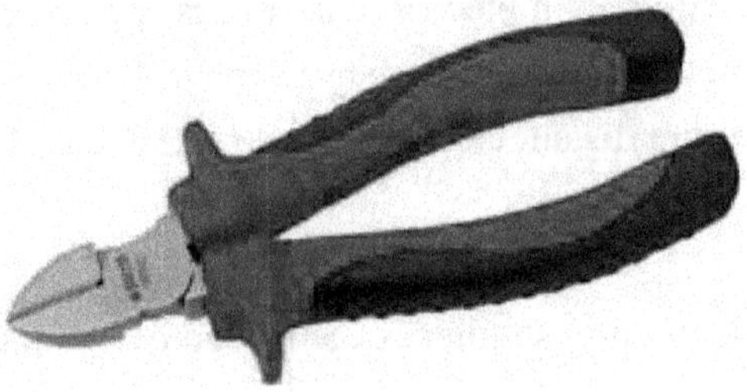

Fig.1-10 Sample of diagonal plier

8) Pressure champ

Pressure clamp, as shown in Figure 1-11, is mainly used for pressing various terminals. The pressure adjusting knob can be used for adjusting the size of the opening nipper to facilitate the use of various terminals. Copper bare-pressed terminals are firmly pressed onto multiple or single wires with cold-pressed clamps.

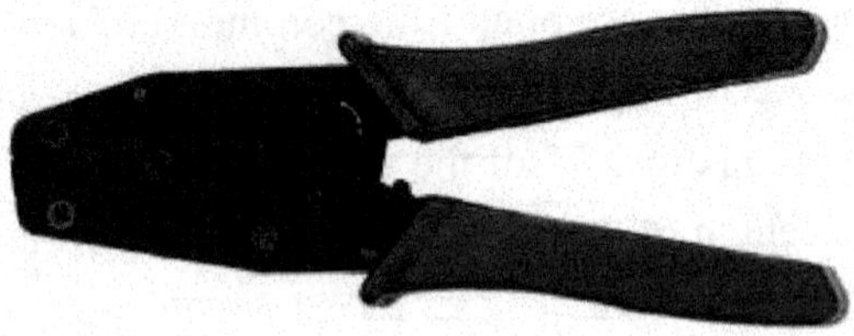

Fig. 1-11 Sample of pressure champ

9) Wire stripper

Wire stripper is one of the commonly used tools for interior line workers, and instrument electricians, as shown in Figure 1-12. It is used by power suppliers to strip the surface insulation layer of the wire head. The stripping clamp can separate the insulating skin of the cut wire from the wire and prevent electric shock.

Fig. 1-12 Stripping plier

Wire stripper is one of the common tools used by internal electricians and instrument electricians to strip the surface insulation layer of wire head.

As shown in Figure 1-12, it consists of pliers and handles. The pliers are composed of wire-pressing pliers and notches. They are divided into several notches of diameters of 0.5-3 mm, which are suitable for the exploitation of wire cores of different specifications.When being used, the cut size should be slightly larger than the diameter of conductor core; otherwise the core wire will be cut off. It is characterized by its convenience. Stripping the insulating layer does not damage the core, and it is suitable for insulating conductors below 6 mm^2. It should be noted that live wire stripping is not allowed during its use.

10) Electrician knife

Electrical cutter is a cutting tool commonly used by electricians, as shown in Figure 1-13. Ordinary electrical knives are composed of blades, knife holders, knife hangers, etc. When not in use, shrink the blade into the handle. The blade root is articulated with the handle, on which there are scale lines and scale marks. The front end forms a screwdriver head. The two sides are machined with the file face area. The blade has a concave curved knife edge. The end of the curved knife edge forms a knife tip, and the knife handle is provided with a protective button to prevent the blade from bouncing back. The blade of the electric knife has many functions. When used, only one electric knife can complete the operation of connecting wires without carrying other tools. It has the advantages of simple structure, convenient use and versatility.

Fig. 1-13 Electrician knife

11) Flat file

A flat file is a hand tool for filing light workpieces, as shown in Figure 1-14. It is used for micro-processing metal, wood, leather and other surface layers.

Fig. 1-14 Flat file

Filing is the operation of cutting workpiece materials with a file. The utility model has a wide range of applications, and can file the outer surface, inner hole, groove and various complicated surfaces of the workpiece. The operation demonstration is shown in Figure 1-15.

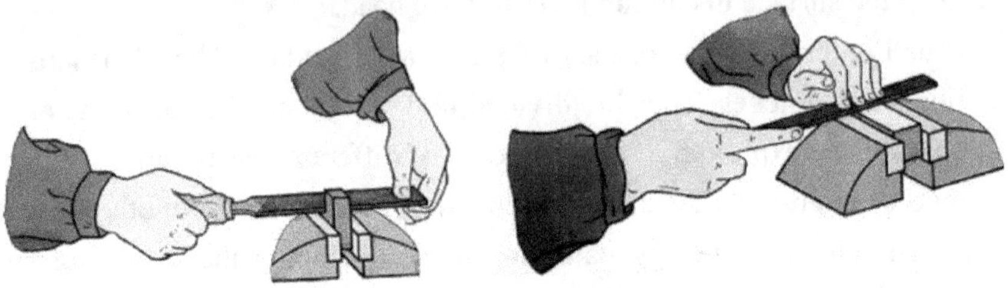

Fig. 1-15　Medium file holding method

The following are the precautions when using the flat file:

(1) Do not hit other objects with a file; the hard and brittle file is easy to break.

(2) Do not use a file without a handle or with a split handle in order to avoid stabbing the wrist.

(3) Do not blow iron chips with your mouth in order to prevent them from flying into your eyes.

(4) Do not touch the surface of the file with your hand in order to prevent slippage during filing.

(5) After the filing surface is blocked, iron filings are brushed in the direction of the tooth pattern with a copper filing brush.

(6)The file should not be put out of the forceps so as not to hit and hurt the foot.

12) Screwdrivers of different shapes

Rotary tool is also called screwdriver. According to its head shape, it can be divided into three types as shown in Figure 1-16.

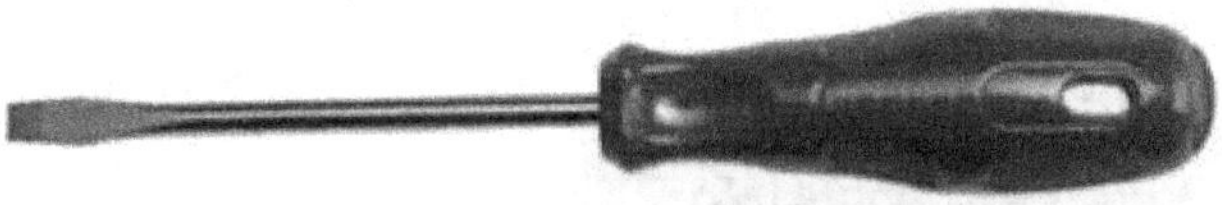

(a) Horizontal-shaped screwdrivers

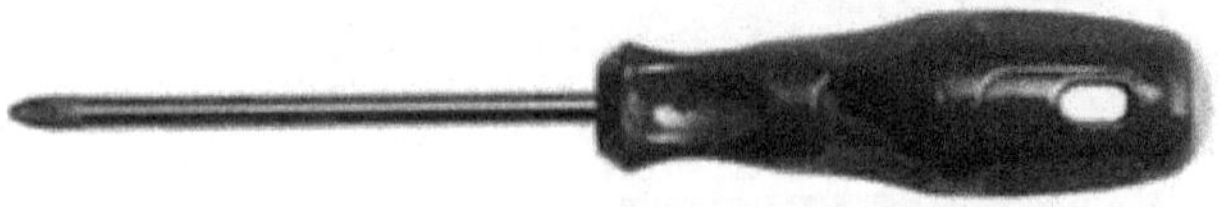

(b) Cross-shaped screwdrivers

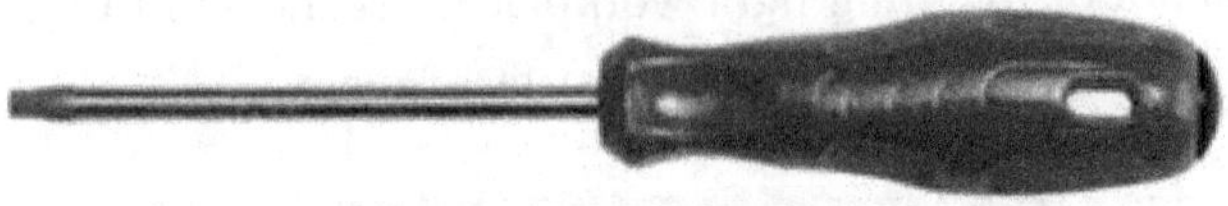

(c) Multiple-shaped screwdrivers

Fig. 1-16　Screwdrivers of different shapes

13) Electric soldering iron (optional tool)

Soldering iron is an essential tool for electronic production and electrical maintenance, as shown in Figure 1-17, which is mainly used to weld components and wires. According to the mechanical structure, it can be divided into internal heating soldering iron and external heating soldering iron; and it can be divided into non tin soldering iron and tin soldering iron in terms of functions; it is divided into high power and small power in terns of power.

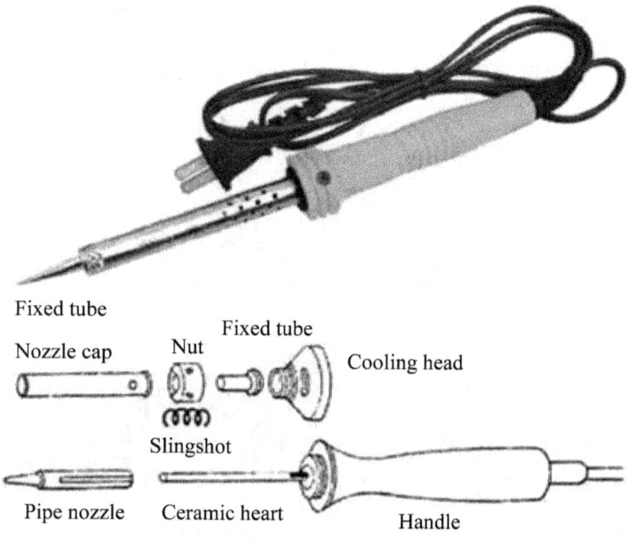

Fig. 1-17 Electric soldering iron

(1) Operation code for electric soldering iron.

Action decomposition formula: preheat soldering iron, melt tin, remove tin wire, and remove soldering iron, cool tin point and solidify tin point.

(2) Standard temperature: The general welding temperature is 265 +5 degrees; ordinary soldering iron (30-60 W) is 260-320 degrees; constant temperature soldering iron is 300-340 degrees.

(3) Basic methods for operation: Soldering iron must be equipped with soldering iron frame, sponge and tin slag box for welding operation. The sponge in the soldering iron rack is immersed in sufficient water to ensure the cleaning of the soldering iron at any time. Insert the soldering iron plug into the power supply, turn on the power switch, and adjust the temperature control switch to the temperature specified in the "Operation Instruction". After about 2-5 minutes of preheating, ask the relevant personnel to measure the temperature of the soldering iron, and then conduct welding operation after all set. Otherwise, the relevant personnel should be allowed to repair or correct the soldering iron. In the process of welding, the soldering head and the soldered joint are generally kept at an angle of 45 degrees, and the soldering head should be in full contact with the soldered object, so that the surface temperature of the soldered object is consistent with that of the solder. If the first welding is not firm, the second welding should be carried out after the temperature of the weld drops completely. The iron head must be cleaned regularly with sponge, and the tin slag should be put into the tin slag box. After the operation, clean the iron head with sponge, then coat it with new tin, then turn off the power button and

pull out the power plug. After the soldering iron is cooled, the dirt on the fuselage and in its surroundings should be cleaned.

(4) Welding requirements:

① Firm connection, tin spot should be perfect and bright, no sharp and burr conditions are allowed.

② There should be no short circuit, cold welding, virtual welding and false welding during welding.

③ Examples of bad welding conditions are shown in Figure 1-18.

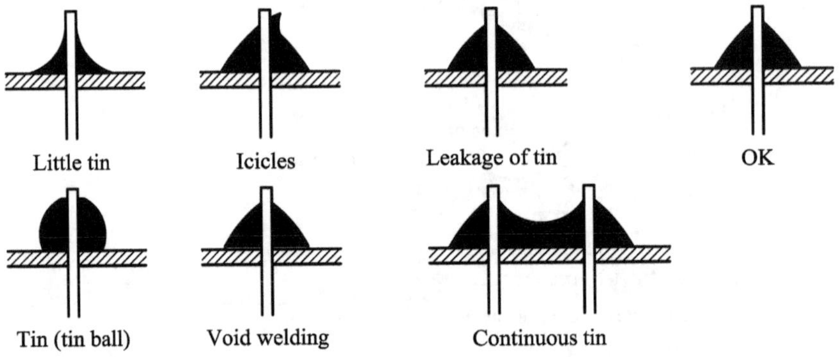

Fig. 1-18　Different welding conditions

(5) Precautions in the use of soldering iron:

① When the power supply is connected, the temperature of the welding head is usually at 200-400 degrees. In view of the possible burns or fires caused by abuse, please strictly observe the following rules: do not touch the metal part that is near the welding head; do not use welded iron head near flammable objects. Welding iron head is extremely hot, which may cause dangerous accidents. Power should be turned off during rest or after completion.

② In order to avoid damaging the electric soldering iron and maintain the safety of the working environment, the following matters should be observed: do not use welded iron joints for work other than welding; do not knock the soldering iron on the worktable to remove flux residue, which may cause serious damage to the soldering iron; do not alter the electric soldering iron without authorization. When replacing parts, hand it to the technician for treatment.

14) Movable wrench

Adjustable spanner for short, as shown in Figure 1-19, whose opening width can be adjusted within a certain range, is a tool used to tighten and loosen nuts and bolts of different specifications. The adjustable spanner is composed of a head and a handle. The head is composed of a movable lip, a mechanical lip, a plate mouth, a turbine and a shaft pin. Rotating turbine can adjust the size of the plate mouth. The specification is expressed as length × maximum opening width (unit: mm). Electricians commonly use the four specifications of 150 mm × 19 mm(6'), 200 mm × 24 mm(8 '), 250 mm × 30 mm(10 ') and 300 mm × 36 mm(12').

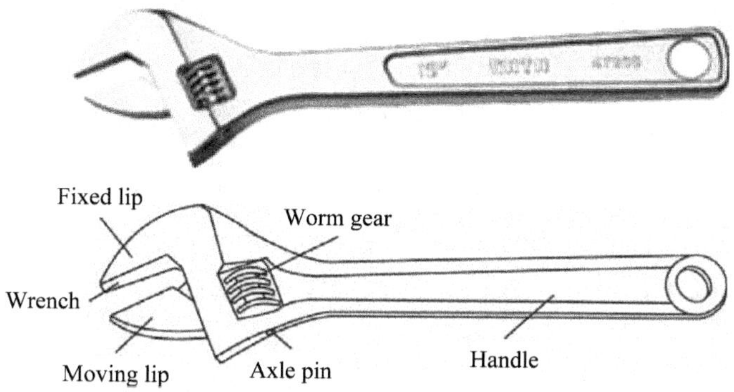

Fig. 1-19 Movable wrench

When using the flexible wrench, the appropriate specification of the flexible wrench should be selected according to the size of the nut. When pulling large nuts, the stronger moment is often used, so the hand should be held at the tail of the handle to increase the moment, which facilitates pulling; when pulling small nuts, the moment needed is not big, but it is easy to slip. Therefore, the hand can be held near the head, and the thumb can be used to adjust and stabilize the screw.

【Practice and Operation】

1. Task preparations

Prepare the necessary materials and equipment according to the experiment content.

In addition to the equipment listed in Table 1-1, other necessities are shown in Table 1-4.

Table1-4 List of the equipment

No.	Name	Unit	Quantity
1	AA battery	Piece	1
2	Button battery	Piece	1
3	Battery of cell phone	Piece	1
4	Newton battery	Piece	1
5	Resistances	Set	Several
6	Light-emitting diode	Set	1
7	Rectifier diode	Set	1

2. Task content

1) Measure the parameters for electrical components with a digital multimeter

(1) Measure DC voltage.

The schematic diagram for measuring voltage with a multimeter is shown in Figure 1-20.

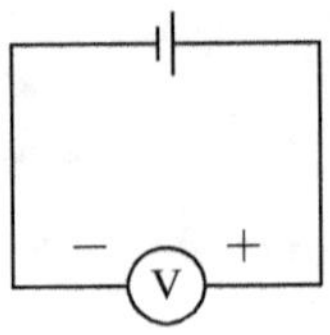

Fig. 1-20 Diagram for voltage measurement

① Measure the voltage of battery No.5. As shown in Figure 1-21, select the appropriate voltage gear, measure the voltage of No. 5 battery as shown in Figure 1-22, and record the measurement data.

Fig. 1-21 Connection diagram

Fig. 1-22 Measure the voltage of battery No.5

② Measure button battery voltage. Select the appropriate gear, and measure the voltage of button battery. The measurement method is shown in Figure 1-23. Record the measurement data.

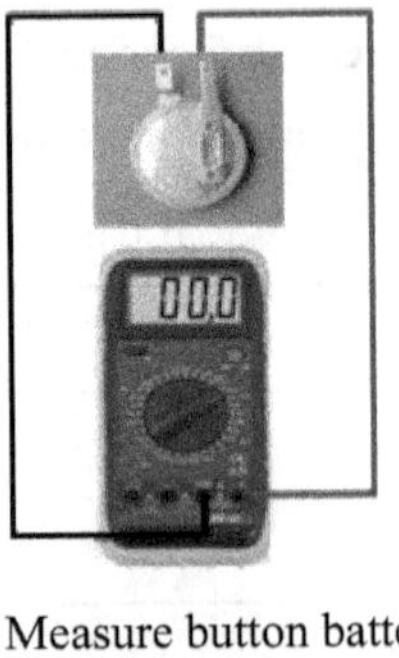

Fig. 1-23 Measure button battery voltage

③ Measure cell phone battery voltage. Take out the battery in the mobile phone, select the appropriate gear, measure its voltage as shown in Figure 1-24, and record the measurement data.

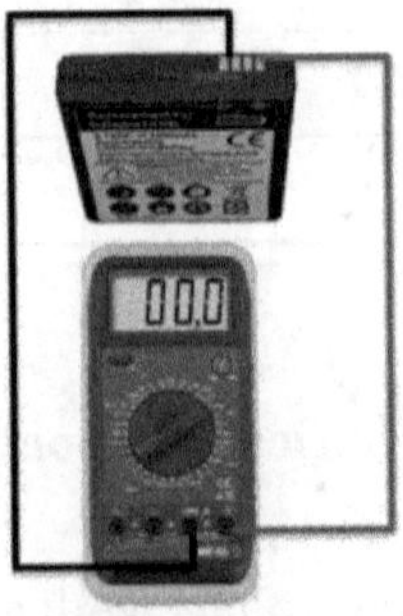

Fig. 1-24 Measure cell phone battery voltage

④ Measure the voltage of carbon battery. Select the appropriate gear, measure the voltage of button battery as shown in Figure 1-25, and record the measurement data.

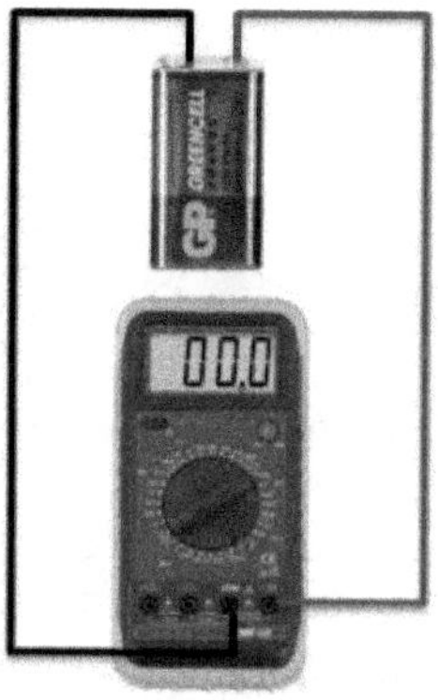

Fig. 1-25 Measure carbon battery voltage

⑤ Measure AC voltage. Insert the multimeter black pen into COM, the red pen into V/Ω, rotate the gear to fit the gear, insert the red and black pen into the fire line and zero line jack of the plugboard as shown in Figure 1-26, and record the measurement data.

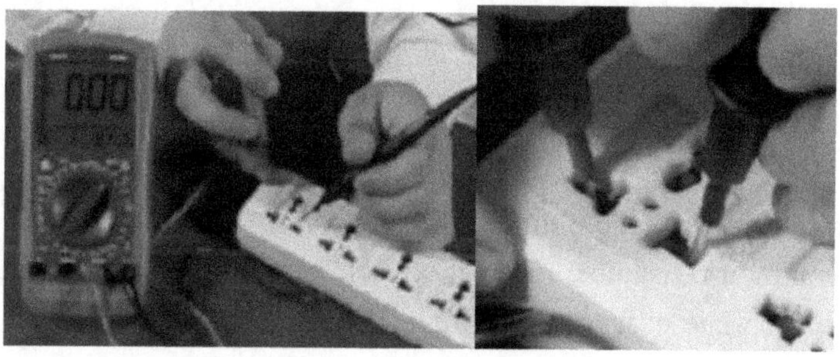

Fig. 1-26 Measure AC voltage

⑥ Measure DC voltage in the circuit. By the connection circuit of Figure 1-27, the parameters for the circuit are measured according to the following lines. R_1 = 10 Ω, R_2 = 1 kΩ, and R_3 = 7.5 kΩ.

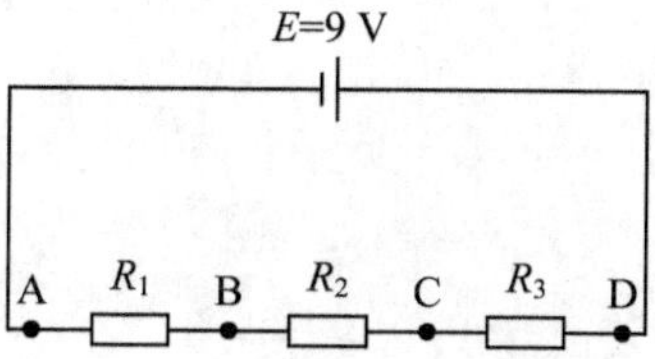

Fig. 1-27 Measure DC voltage

(2) Measure current with a multimeter.

① Measure DC current. The schematic diagram for DC current measurement connection is shown in Figure 1-28, and the measurement method is shown in Figure 1-29. The resistance of R_1 in the circuit is replaced, the current in the circuit is measured, and record the measurement data.

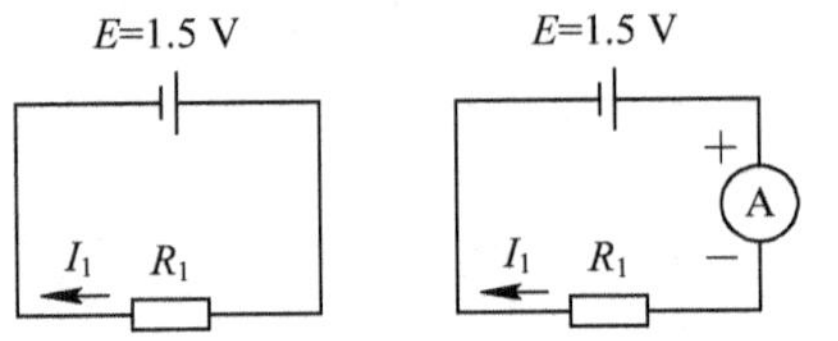

Fig. 1-28 Diagram for measuring DC current

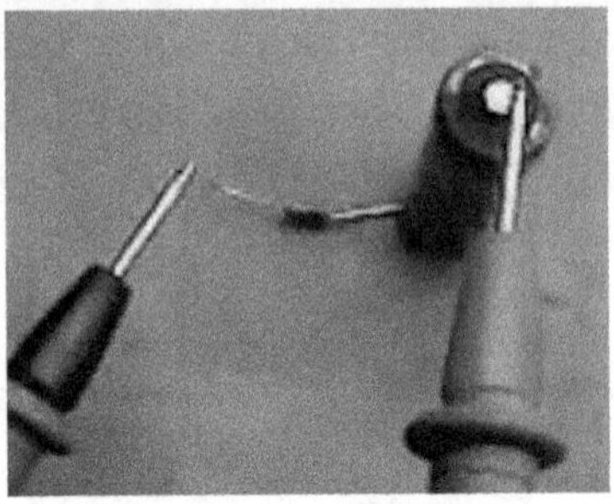

Fig. 1-29 Measure DC Current

② Measure AC current.

The schematic diagram for AC current measurement connection is shown in Figure 1-30. Replace the R_2 resistance value in the circuit, measure the current value in the circuit, and record the measurement data.

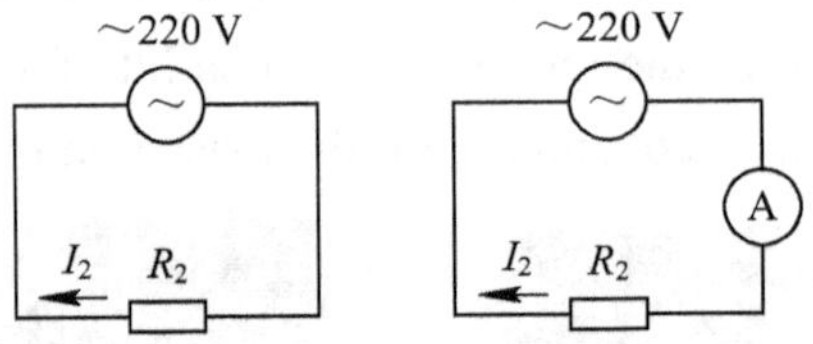

Fig. 1-30 Measure AC Current

(3) Measure resistance with a multimeter.

① Measure common resistance. Insert the black pen into "COM", insert the red pen into "V/Ω", select the appropriate gear, measure the resistance values as shown in Figure 1-31, and calculate the relative error values.

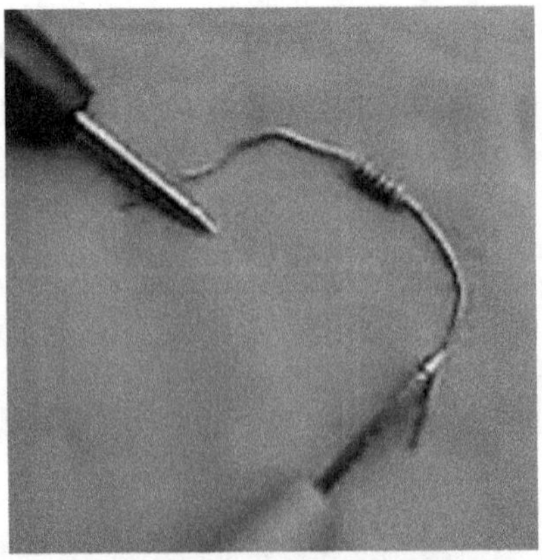

Fig. 1-31 Measure common resistance

② Measure adjustable resistance. Insert the black pen into "COM" and the red pen into"V/Ω", select the appropriate gear, measure the resistance values as shown in Figure 1-27, and calculate the relative error values.

Interesting experiment: Do you know the resistance value of your body?

Let's measure the resistance of our body.

As shown in Figure 1-32, the resistance of each group member's body is measured. Record the resistance value.

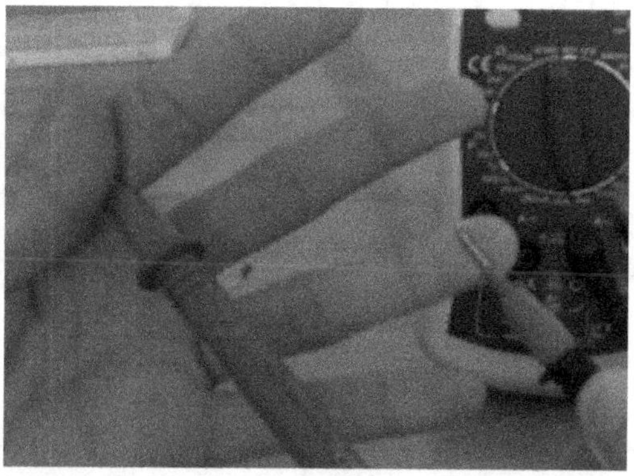

Fig. 1-32 Measure resistance of our body

(4) Continuous conduction test of diode and buzz.

① Detection of positive and negative electrodes of light emitting diodes are shown in following steps:

- Place the digital multimeter in the on-voltage drop detection gear (diode gear) as shown in Figure 1-33.
- Connect the red watch pen to one end of the light emitting diode, and the black watch pen to the other end.
- If the diode emits light and the multimeter display screen shows a conduction voltage drop of about 1.744 or 1.799, the red pen is connected to the positive electrode and the black pen is connected to the negative electrode.

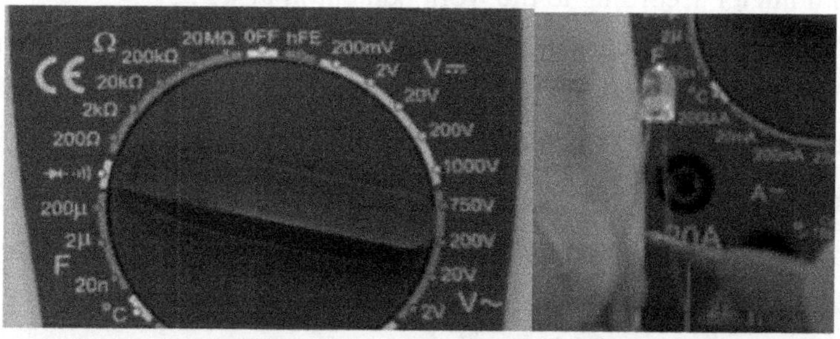

Fig. 1-33 Continuous conduction test of diode and buzz

② Detection of leakage of common rectifier diodes is shown in following steps:

- Place the multimeter in the PN junction voltage drop detection file (diode file).
- Connect the red pen to the one end of the diode and the black pen to the other end. If the digital multimeter shows an on-off voltage drop of about 0.581, the red pen is connected to the positive and the black pen is connected to the negative.

2) Installation, wiring and operation of indoor power and lighting control circuits

The lighting control circuit is shown in Figure 1-34.

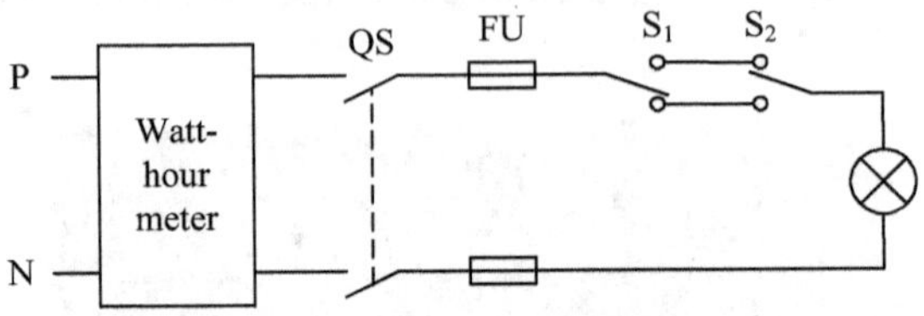

Fig. 1-34 Lighting control circuit

The experiment can be carried out according to Figure 1-35 only after the instructor checks the connection correctly.

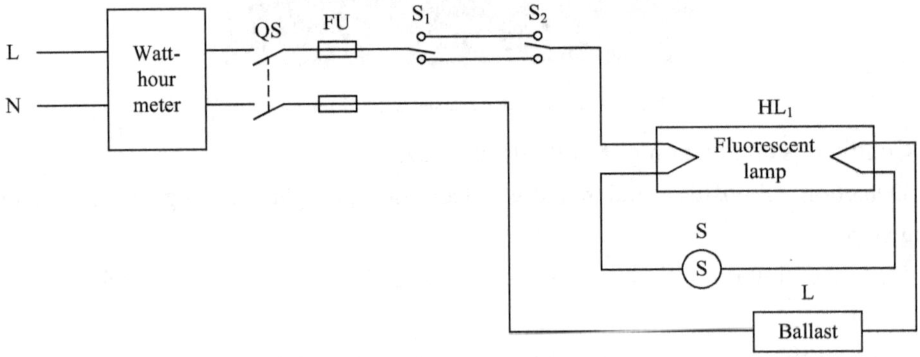

Fig. 1-35 Fluorescent lamp control circuit

3. Task procedures

(1) Use the multimeter to measure the parameters for relevant electrical components, and learn to use the multimeter according to the work content in task 1.

(2) Complete the wiring according to the control circuit diagram for fluorescent lamp in Figure 1-35, pay attention to the wiring process and the arrangement of electrical components on the training wiring hanging board, and make it standard and orderly.

4. Precautions

Once the wiring has been checked for correctness by the instructor, the power must be turned on under the supervision of the instructor.

5. Notes and summary

____

____

____

____

Task 2 Installation, Connection and Operation of the Standby Power Supply Circuit for Automatic Switching of Overcurrent in Power Grid

【Relevant Knowledge】

The dual power supply is generally composed of a main power supply and a standby power supply, in order to be able to switch to the standby circuit power supply in case of a sudden main circuit power failure. Among them, the dual power supply automatic switch is a microprocessor-controlled device for automatically switching between grid power and grid power or grid power and generator power in the grid system to provide continuous power supply. When the common power suddenly fails or power outage occurs, the dual power automatic switch can be automatically put into the backup power (the backup power under small load can also be powered by the generator), so that the equipment can still operate normally. The most common applications are elevators, firefighting, monitoring, lighting and so on.

The development of dual power switching system products broadly evolved into three types of products: contactors, molded case circuit breakers / load disconnect switches, and integrated automatic transfer switch electrical appliances.

1. Contactor

This kind of power switching system uses a contactor as the switching actuator, and the switching function is achieved by using an intermediate relay or logic control module to form a secondary circuit. The disadvantage is that the work of the main circuit contactor requires the secondary circuit to be energized for a long time, which is likely to cause a temperature rise, fever, contact bonding, coil burnout and other faults. A non-standard product, it consists of many components. The product is poor in quality due to its components and manufacturing process, resulting in a high failure rate, and thus it has been gradually replaced by new products.

2. Plastic case circuit breaker

This kind of power switching system uses molded case circuit breaker as the switching control part, and the switching function is achieved by an ATS automatic control unit. There are mechanical and electrical power supply. It has the advantages of perfect function, good operating performance, long service life, fewer components and easy installation. This type is a CB level changeover switch, which consists of two circuit breakers as current breaking unit and which is equipped with current disconnector and has a certain protective capability. Circuit breakers have a much higher ON/OFF capability than relays.

These products are maintained by the mechanical structure during steady state, but due to the difference between the circuit breaker and load disconnect switch, they are not as effective as PC level products in overcurrent condition.

3. Load disconnect switch

The type of power switching system is based on two load disconnect switches with an electric operation mechanism, a mechanical interlocking mechanism, and an automatic control unit assembled as an integrated unit. The current breaking unit is a load disconnect switch, and its contact arc extinguishing system is designed to meet the requirement for breaking the primary arc, without the circuit protection function. This category of products are PC-level products that can be quickly switched on due to the spring-loaded energy storage and instantaneous release acceleration mechanism. The products are reliable in operation performance, such as breaking the circuit or converting the circuit.

4. Integrated automatic transfer switch

This type of power switching system is a combination of switching and logical control, without the need for an external controller, to achieve real mechatronics automatic transfer switch. The contact system of such power switching system products adopts the “single-pole double-throw” design, which is uniformly designed and manufactured to be of small size and simple structure. The product does not have a current protection function, a PC-level transfer switch electrical product. The conversion time of such products is relatively short, motor drive is used as the switch-drive, switching is smooth and reliable, the motor drive operator has current passing only during the switching and there is no need to provide working current in the steady state, which is remarkably energy-saving. The product shows no signs of temperature rise fever, contact bonding and coil burning. The switch is equipped with mechatronics interlocking device, which can realize self-recovery, manual-automatic conversion and delay-control, and which can also realize drop voltage, low voltage and phase break protection. It is a mainstream product for power switching.

【Practice and Operation】

1. Task preparations

In addition to the devices that are listed in Table 1-1, the equipment required for this task is shown in Table 1-5.

Table 1-5　Equipment for skills training

Code	Name	Mode	Quantity
QS	Three-phase leakage switch	DZ47LE-32	1
FU	Fuse	RL1-15	2
SB_1	Start button	LA19-11	1
KM	AC contactor	CJ20-10/220V	1
L	Ballast		1
S	Starter		1

Continued

Code	Name	Mode	Quantity
HL_1	Indicator light	AD11-25/40	2
K	Dual controlled switch		2
KA	Overcurrent relay	JL14	1
W	Single-phase watt-hour meter	DD282	1
HL_2	Fluorescent lamp	8W	1
	Terminals, slots, conductors		Suitable numbers

2. Task content

The schematic diagram for automatic switching standby power supply is shown in Figure 1-36. The power is introduced into the watt-hour meter by pressing the button SB_1, KT coil, current relay coil and KM coil. The auxiliary contacts are open when KT power off is delayed, and the normal closed contacts of KM are disconnected to prevent the standby power supply from supplying power. KM usually opens auxiliary contacts and closes self-locking.

When the external main power supply is cut off and the standby power supply is started, the indicator HL_1 lights up. It can be seen from the circuit that S_1 and S_2 are two different switches. Whichever switch is closed or disconnected, the fluorescent lamp can be controlled to being turning on or off, so as to control the two places. The overcurrent relay KA in the figure is mainly for overcurrent protection. When the current in the circuit is too large, it will automatically cut off the power supply.

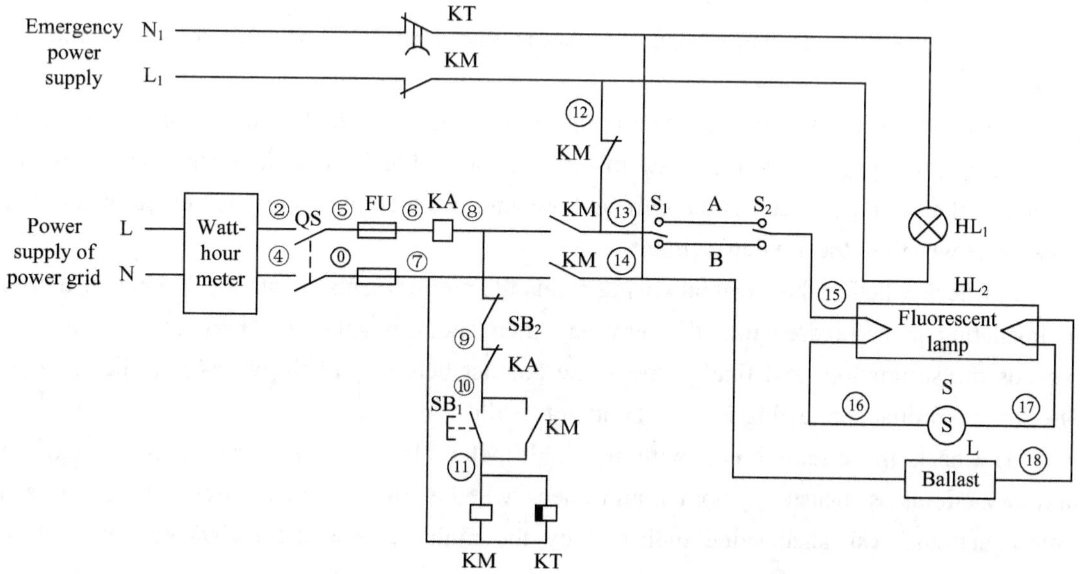

Fig. 1-36 Automatic switching backup power supply for overcurrent in power grid

3. Task procedures

The installation wiring diagram for the automatic switching standby power supply circuit is shown in Figure 1-37. The required length is measured between fixed electrical components based on the wire number of the wiring diagram, and the appropriate length of the wire is intercepted. Remove the insulating layer at both ends of the conductor and insert the conductor into the terminal directly after the line number tube is inserted.

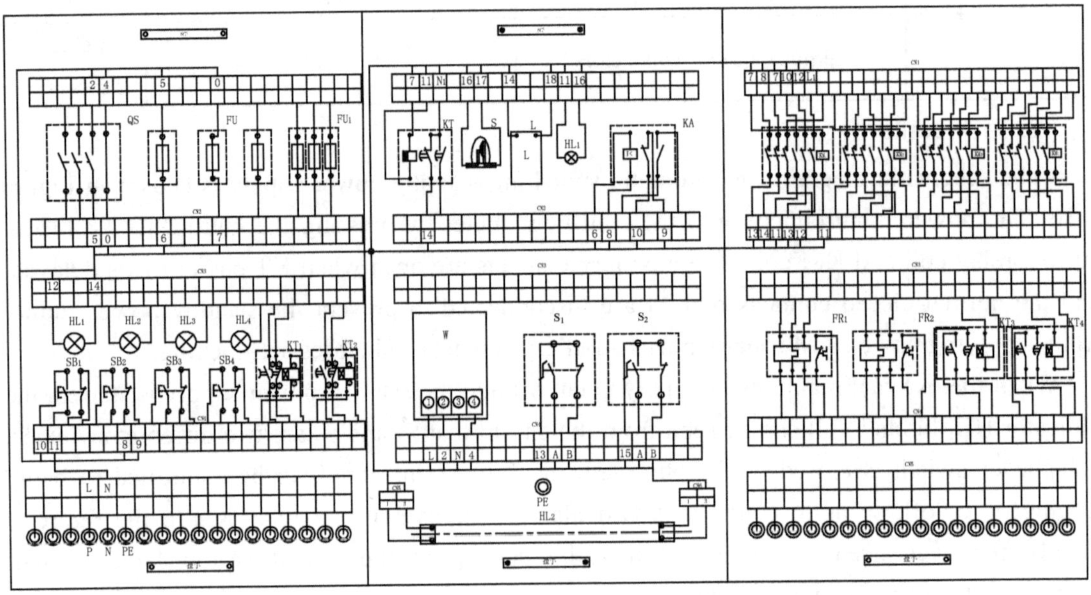

Fig. 1-37 Installation wiring diagram for automatic switching standby power supply circuit

After connecting the equipment, we should check the line mainly from the following aspects:

(1) Check wiring. After the connection is completed, check the terminal connection number from the power supply end to eliminate misconnection and leakage, and check whether the line number of the same conductor is the same at both ends, with emphasis on the line numbers likely to be made wrong in the auxiliary circuit.

(2) Check whether the terminal wiring meets the requirements. First, check whether there is an insulating layer pressed into the terminal, then check whether the bare core of the wire exceeds the stipulation, and finally check the contact between all the wires and the terminal. Check the terminal connecting wire by hand not to allow any loose.

(3) Check the circuit break with the multimeter "R×10". First, take two fuses of the auxiliary circuit as measuring points, and check whether there is short circuit. Press the start button, and the resistance value indicated by the multimeter is the resistance value of the contactor coil, indicating that the auxiliary circuit is connected when the start button is pressed. Then press the touching point of the contactor, and the resistance value indicated by the multimeter is the direct current resistance of the electromagnetic coil of the contactor, indicating

that the auxiliary circuit can self-lock. Press the start button and when the table pointer is deflected, press the stop button. The table pointer returns to "∞", indicating that the stop button can disconnect the auxiliary circuit.

4. Precautions

After checking the connection, test must be carried out under the supervision of the teacher.

5. Notes and summary

【Extension Exercises】

Exercise 1 Laying of Various Pipes

1. Task objectives

(1) Familiarize students with all kinds of electrical tools.

(2) Master the laying process of bridge shelf, PVC pipe, trunking, metal pipe, etc.

2. Equipment

The devices needed are bevel pliers, manual pipe bender, pipe bender spring, steel straightedge, steel tape measure, angle ruler, hand saw bow, saw blade, hand hammer, drill bit, etc. In addition, a variety of other equipment required is is shown in Table 1-6.

Table 1-6 Equipment for pipe laying line

No.	Name	Specification	Quantity
1	PVC tube	Φ16	1 meter
2	PVC cup drain	Φ16	1
3	PVC cup drain	Φ25	1
4	Spike card	Φ16	3
5	Metal tube	Φ25	0.4 meters
6	Trunking	Φ25	0.6 meters
7	Trunking	Φ50	1.5 meters
8	Bridging	Φ100	1
9	86 Switch box		3
10	Sheath wire		0.5 meters
11	Sheath wire card		3
12	Waterproof joint		1

3. Task contents and procedures

(1) The diagram for pipe laying line is shown in Figure 1-38.

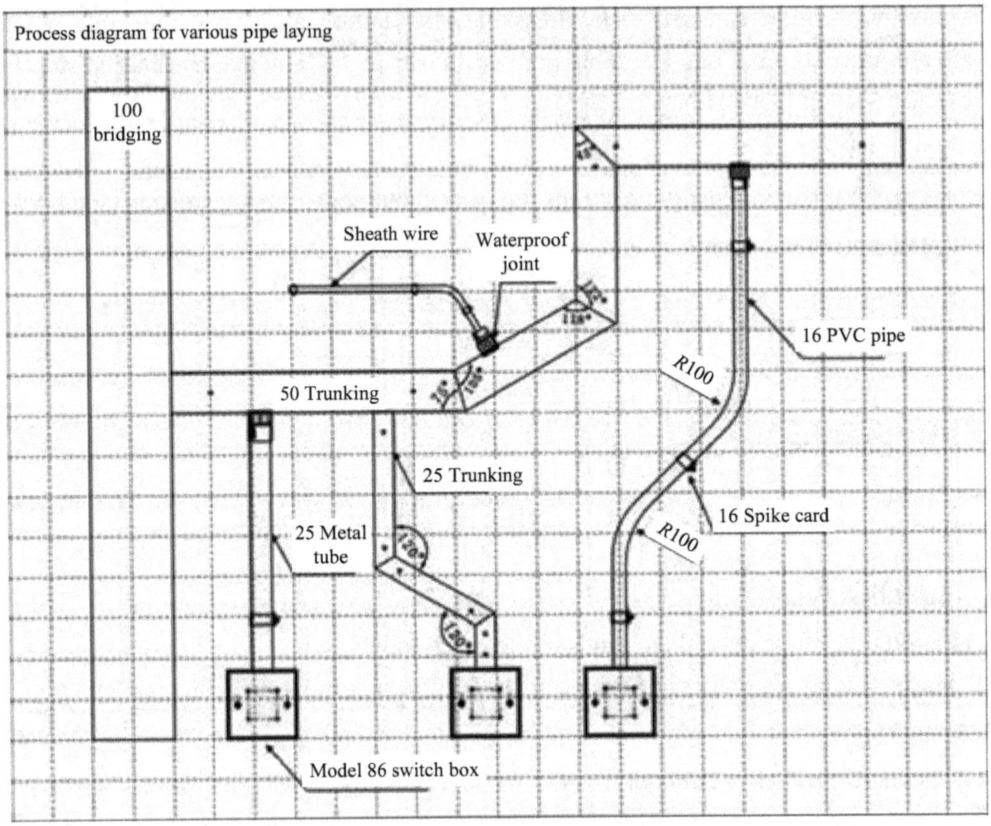

Fig. 1-38　Diagram for pipe laying line

(2) Pipeline prefabrication and processing: PVC pipes shall be processed into required bending degree and length; trunking shall be processed into required angle and length; sheath wires shall be cut into required length.

(3) Measure the position of 86 switch box: measure the exact position of 86 type switch box.

(4) Pipeline connection: connect all kinds of processed pipes by the drawing.

(5) Pipeline fixing: fix the pipeline with the supporting fixing equipment for various pipes.

Be aware of the following steps when conducting the task:

(1) The interface between the switch box and the wire duct should be set properly.

(2) The location and spacing of various fixed equipment shall be calculated in advance and reasonably arranged.

(3) When fixing the pipe clamp, make sure that it is horizontal or vertical, without tilting.

Exercise 2　Installation of Voltmeter and Ammeter Circuits

1. Task objectives

(1) Master the working principle of voltmeter and ammeter.

(2) Learn the installation and wiring of voltmeter and ammeter.

2. Equipment

The electrical components required for this training exercise are shown in Table 1-7.

Table 1-7 Required electrical materials

No.	Name	Specification	Quantity	Note
1	Leakage protector	DZ47LE-2P-10A	1	Q
2	Screw socket		1	
3	Spiral incandescent lamp	220 V/25 W	1	
4	AC voltmeter	Type 6L2 voltmeter	1	
5	AC ammeter	Type 6L2 ammeter	1	
6	Resistance	75 Ω/75 W	1	R_1

3. Task contents and procedures

(1) Master the wiring principle of voltmeter and ammeter as shown in Figure 1-39. In the figure, Z is the load of the circuit.

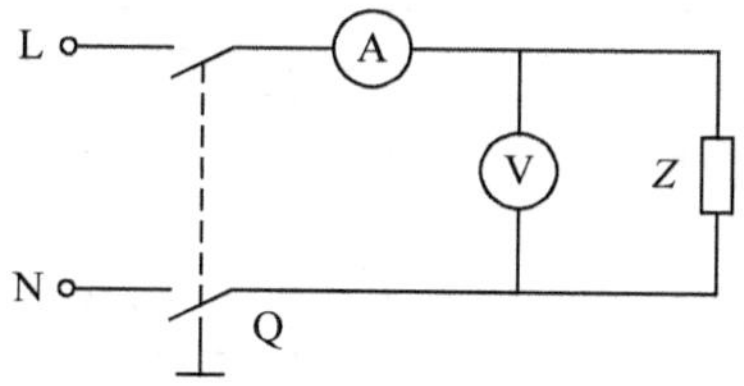

Fig. 1-39 Wiring schematic diagram for voltmeter and ammeter

(2) Installation and connection.

The incandescent lamp is used as the load in this training, and the wiring is as shown in Figure 1-40. Note: because the bulb power is small, the range of ammeter is large, and the current indication is not obvious, the resistance R_1 can be used as the load.

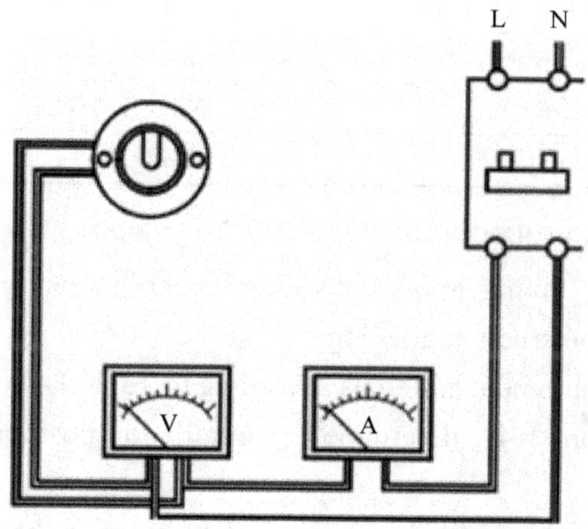

Fig. 1-40 Wiring diagram for voltmeter and ammeter installation circuit

(3) Testing and debugging.

After checking that the wiring is correct, press the start button on the control panel to start the power supply, close the switch Q, the load bulb is on, and observe the indication of the voltmeter and ammeter.

Exercise 3 Installation of Energy Saving Lamp and Socket

1. Task objectives

(1) Familiarize students with all kinds of electrical tools.

(2) Master the installation and wiring of energy-saving lamps and sockets.

2. Equipment

The devices needed are diagonal pliers, manual bender, bending spring, steel ruler, steel tape, angle ruler, hand saw bow, saw blade, hammer, electric hand drill, drill, screwdriver, etc. And other devices are shown in Table 1-8.

Table 1-8 Equipment for energy saving lamp and socket line

No.	Name	Specification	Quantity
1	PVC tube	Φ25	3 meters
2	PVC tube	Φ16	1 meter
3	PVC cup drain	Φ25	5
4	PVC cup drain	Φ16	2
5	Type 86 switch box		3
6	Power distribution box		1
7	Leakage protector	DZ47LE-2P-25A	1
8	Air switch	DZ47-1P-3A	2
9	Screw flat lamp		1
10	Energy saving light bulb	220 V/8 W	1
11	Round wood		1
12	Voice switch	CD200-D86SG	1
13	Power supply socket	DG862K1	2

3. Task contents and procedures

Install and wire energy-saving lamps and sockets according to Figure 1-41.

(1) Be familiar with construction drawings.

(2) Prepare relevant equipment and tools according to Table 1-8.

(3) According to Figure 1-41, determine the installation position of electrical appliances, laying route of wires, etc.

(4) All fixing points shall be drilled with installation holes on the training hanging plate.

(5) Install pipe clamp, PVC pipe and various mounting brackets.

(6) Laying wires: lay wires according to the schematic diagram shown in Figure 1-42.

(7) Installation of lamps and electrical appliances: fix the bulbs and switch socket panels.

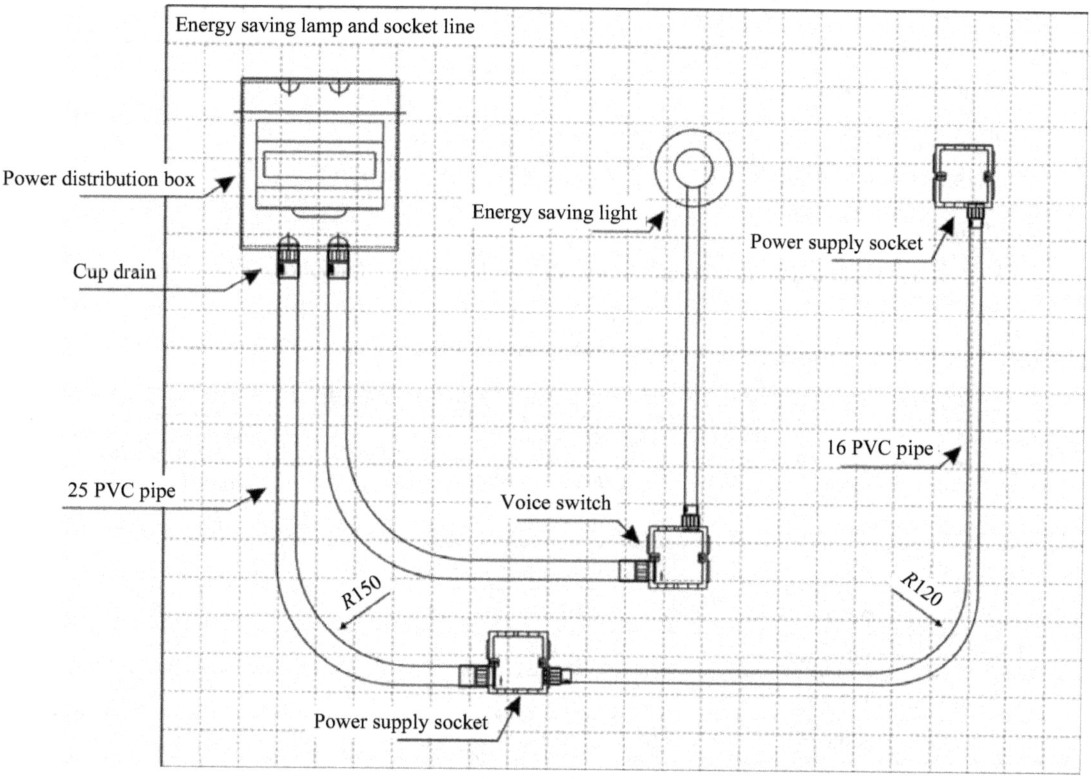

Fig. 1-41 Installation and connection diagram for energy-saving lamps and sockets

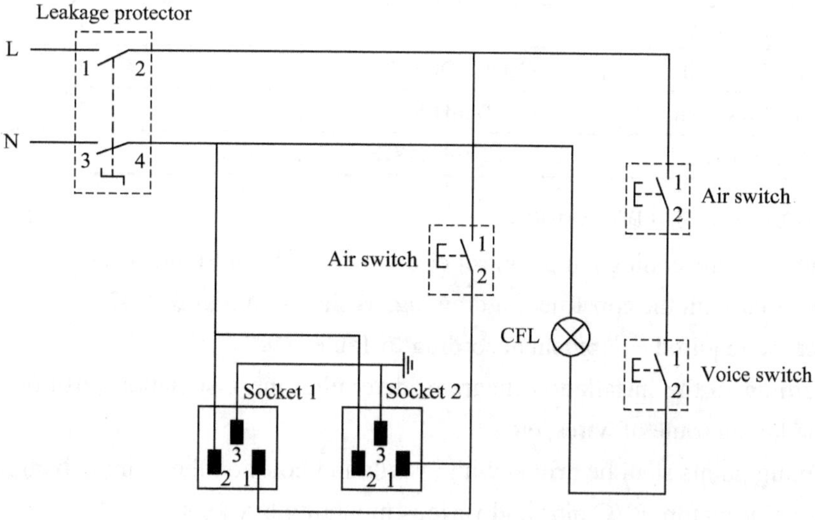

Fig. 1-42 Schematic diagram for energy-saving lamps and sockets

Exercise 4　Installation of Ceiling Lamp Control Circuit

1. Task objectives

(1) Familiarize students with all kinds of electrical tools.

(2) Master the installation and wiring of ceiling lamp control circuit.

2. Equipment

The devices needed are diagonal pliers, manual bender, bending spring, steel ruler, steel tape, angle ruler, hand saw bow, saw blade, hammer, electric hand drill, drill, screwdriver, etc. And the electrical components are listed in Table 1-9.

Table 1-9　List of electrical components of ceiling lamp control circuit

No.	Name	Specification	Quantity
1	PVC tube	Φ25	3 meters
2	PVC tube	Φ16	1 meter
3	PVC cup drain	Φ25	7
4	PVC cup drain	Φ16	1
5	Type 86 switch box		2
6	Power distribution box		1
7	Leakage protector	DZ47-LE-2P-25A	1
8	Air switch	DZ47-1P-3A	1
9	Screw flat lamp holder		1
10	Incandescent light bulb	220 V/40 W	1
11	Round wood		1
12	Single switch	CD200-DG86K2	1
13	Infrared switch	D86HW	1
14	Voice switch	CD200-D86SG	1

3. Task contents and procedures

Install and wire the ceiling lamp control circuit according to Figure 1-43.

(1) Be familiar with the construction drawing, as shown in Figure 1-43.

(2) Select the required equipment according to Table 1-9.

(3) According to the installation diagram, determine the installation position of electrical appliances, the laying route of wires, etc.

(4) All fixing points shall be drilled with installation holes on the training hanging plate.

(5) Install pipe clamp, PVC pipe and various mounting brackets.

(6) Laying wires: lay wires according to Figure 1-44.

(7) Fix the bulbs and switch panels.

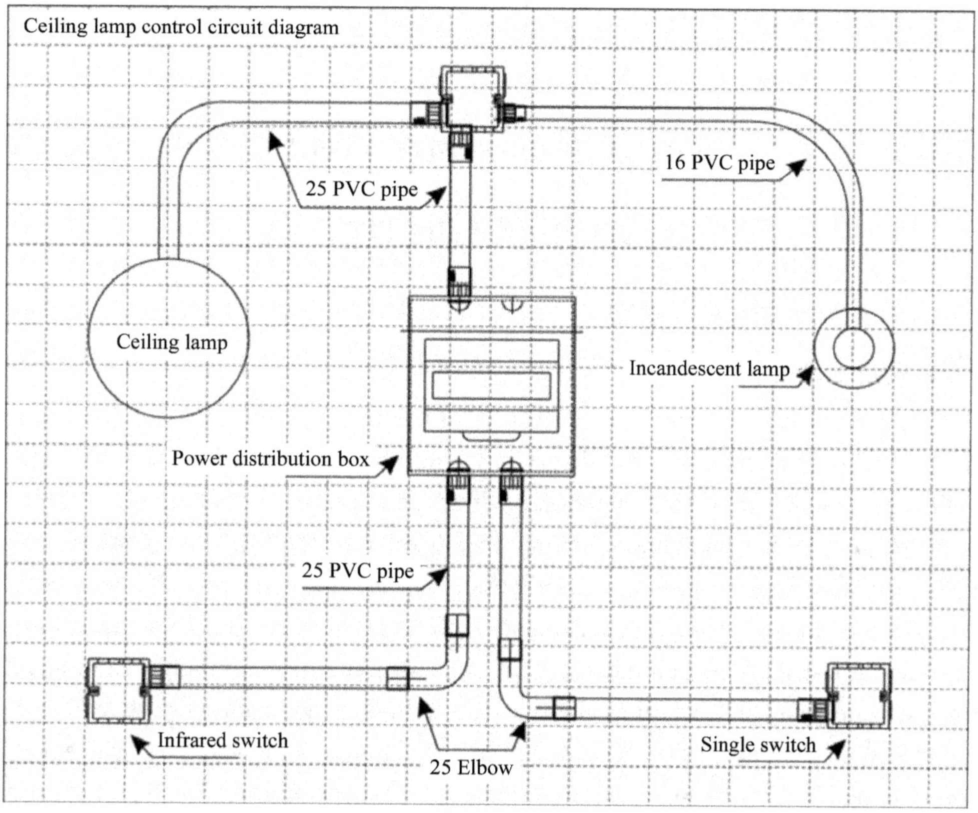

Fig. 1-43 Installation diagram for ceiling lamp control circuit

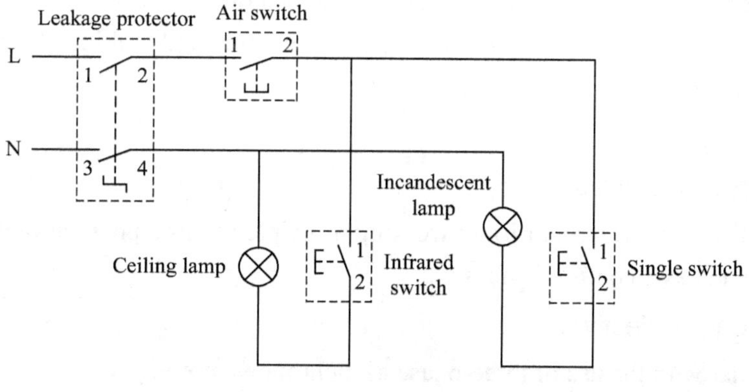

Fig. 1-44 Schematic diagram for ceiling lamp control circuit

Module Two Comprehensive Training in Electrical Control Circuit

【General Introduction】

The comprehensive training in electrical control circuit mainly includes four tasks: the installation, wiring and operation of direct start control circuit of three-phase asynchronous motor; the installation, wiring and operation of direct and reverse control circuit of three-phase asynchronous motor; the installation, wiring and operation of Y-△ start control circuit of three-phase asynchronous motor; the installation, wiring and operation of direct switching control circuit of two-speed three-phase asynchronous motor. Before getting started with the practical training tasks, you should have a brief understanding of the training requirements and related knowledge, and carefully study the practical training tasks that you have learned before. You should also be clear about the objectives of the hands-on training and the issues noted in the hands-on training. Make sure that you know the general scope and trend of the data measured during the training, and prepare the forms for recording the data according to the training contents. It is necessary to understand the following aspects: ①basic structure; ②working principle; ③mechanical characteristics indicating the relationship between speed and torque; ④basic principles of and methods for starting, speed regulating and braking; ⑤applications and how to use them correctly.

【Basic Elements】

1. Knowledge objectives

(1) Be familiar with the structure and working principle of three-phase asynchronous motor.

(2) Understand basic control devices.

2. Competence objectives

(1) Be familiar with the use of three-phase asynchronous motor.

(2) Make the start-up and speed regulation analysis of asynchronous motor.

(3) Master the correct wiring method and operation method for motor forward and reverse control circuit.

3. Operation standards

(1) The experiment is carried out in a group of two to three persons, each group consisting of two to three persons. There should be a clear division of labor concerning the connection, load adjustment, measurement and data recording to ensure the coordination in the operation of the

experiment and the accuracy and reliability of the recorded data.

(2) Before the experiment, it is necessary to be familiar with the components used in the experiment and record the nameplate data of the motor.

(3) Connection must be made according to the experimental circuit diagram, which needs to be simple and clear, and which should be reasonable in layout, and which should be convenient in operation. By the wiring principle, the main circuit in series should be connected first, and then the branches in parallel.

(4) Start the motor based on certain specifications. And observe whether all the instruments are normal (such as whether the forward and reverse direction of the pointer exceeds the full range, etc.). If there is any abnormality, the power supply should be cut off immediately and the fault should be eliminated; if everything is normal, the experiment can be officially started.

(5) According to the operation methods and steps presented in the experimental textbooks, complete the operation process.

(6) When the experiment is completed, disconnection can be carried out upon the instructor's approval.

4. Safety standards

In order to complete the basic experiment of motor and drive on time and ensure the safety of the personnel and equipment during the experiment, the following safety operation procedures should be strictly observed:

(1) The total power supply of the laboratory is operated and controlled by the experimental instructor, and no other personnel are allowed to operate it.

(2) Attention should be paid to preventing clothes, scarves, braids and experimental wires from being involved in the rotating parts of the motor.

(3) Wiring or disconnection must be carried out under the condition of cutting off the power supply. It is strictly forbidden to contact the bare part of the live line and the rotating parts of the motor.

(4) Students must be supervised and allowed by the instructor after completing the wiring or changing the wiring independently, and then remind other students in the group to take care before they can turn on the power supply. If an accident occurs in the experiment, the power supply should be cut off immediately, and the experiment can be continued only after the problems are identified and the faults are properly handled.

(5) If the motor starts directly, the current range of the power meter and ammeter should be checked before starting.

(6) After the experiment, the power supply should be cut off, the experimental wires should be arranged, and the experimental table should be cleaned up.

5. Quality standards

(1) The wiring of the control circuit should be correct, reasonable and beautiful.

(2) According to the measured data, the phenomena observed in the training and the

problems found, an analysis and discussion are made, and a reasonable conclusion is drawn.

【Training Requirements】

1. Preparation of tools, instruments and equipment

Before proceeding with the experiment, prepare a list of all equipment configurations in Table 2-1.

Table 2-1　List of laboratory equipment

No.	Name	Unit	quantity
1	Yalong YL-WXD series training platform	Set	1
2	Hanging plate 1	Piece	1
3	Hanging plate 2	Piece	1
4	Three-phase asynchronous motor	Set	1
5	Auxiliary tools	Piece	Several

2. Knowledge assessment points

(1) Master the principle of motor starting, forward and reverse control circuit.

(2) Master the basic working principle of motor and the basic methods for starting, braking and speed regulating.

3. Skills assessment points

(1) Through the learning and training of schematic diagram in the practical training task, acquire the basic ability to read electrical diagram.

(2) Through the installation, wiring and operation of the three-phase asynchronous motor direct start control circuit, master and deepen the understanding of the wiring method for its control circuit.

(3) Through the installation, wiring and operation of the three-phase asynchronous motor forward and reverse control circuit, master the actual operation skills of wiring according to the schematic diagram.

(4) Through the installation, wiring and operation of Y-△ start control circuit of three-phase asynchronous motor, further understand the principle of Y-△ step-down start.

【Work Tasks】

Task 1　Installation, Connection and Operation of Direct Start Control Circuit of Three-Phase Asynchronous Motor

【Relevant Knowledge】

1. Three-phase asynchronous motor

1) Overview of motors

Motor is the power equipment that converts electric energy into mechanical energy and outputs mechanical torque.

General motors can be divided into DC motor and AC motor. AC motor can be divided into single-phase motor and three-phase motor according to the number of power supply phases. Three-phase motor can be divided into synchronous motor and asynchronous motor. Asynchronous motor can be divided into cage motor and winding motor in terms of different rotor structures.

The three-phase asynchronous motor is simple in structure, convenient in maintenance and reliable in operation. Therefore, it is widely used to do medium and small rolling mills, various machine tools as well as light industrial machinery and draught parts of blowers. According to statistics, about 90% of the domestic electric drive machinery uses asynchronous motors, of which, small asynchronous motors account for more than 70%. Asynchronous motors account for more than 60% of the total load in the power grid.

2) Basic structure of three-phase cage asynchronous motor

Three-phase cage asynchronous motor is mainly composed of stator and rotor, as shown in Figure 2-1.

The stator part of the three-phase asynchronous motor includes the stand, the stator core and the stator winding. The stand is made of cast iron or cast steel. It supports the stator core. The stator core is made of overlapping silicon steel sheets insulated from each other, with slots in the inner circle and stator windings embedded in the slots, as shown in Figure 2-2.

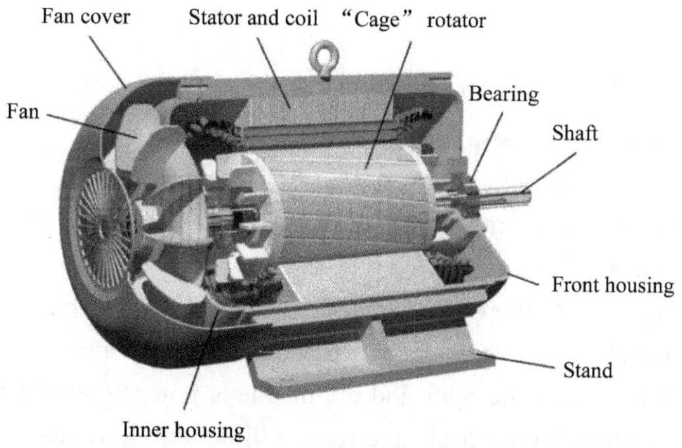

Fig. 2-1 Structural diagram for three-phase cage asynchronous motor

The stator winding is the circuit part of the stator, which consists of three-phase symmetrical windings. Each phase of the three-phase windings is independent of each other.It is embedded in the stator slot at an electrical angle of 120° different from each other, and is insulated from the stator core. The first end of the stator winding is represented by U_1, V_1 and W_1 respectively, while the other end of the stator winding is represented by U_2, V_2 and W_2 respectively.

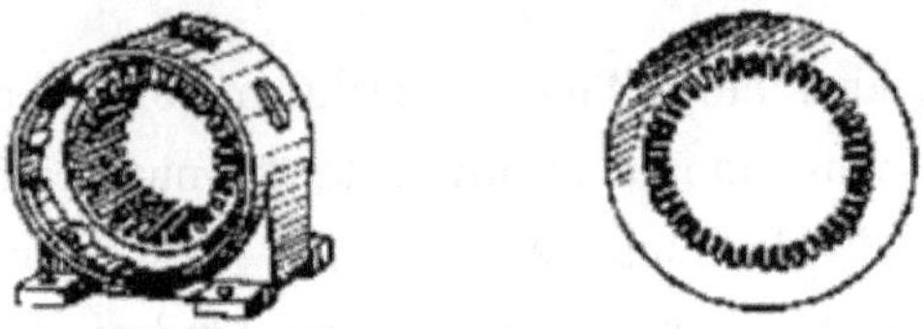

(a) Stator core　　(b) Electronic core punch

Fig. 2-2　Stator core structure

The rotor consists of rotor core, rotor winding and rotating shaft. The rotor core is made of silicon steel sheets with grooves in the outer circle, and copper bars (or cast aluminum) are placed in the grooves. The two ends of the core are connected by conductive end rings to form a short circuit. If the core of the rotor is removed, the structure of the rotor is similar to that of the cage, as shown in Figure 2-3.

Winding asynchronous motor is different only in rotor structure. Its rotor is composed of windings, which are also three-phase like stator windings.

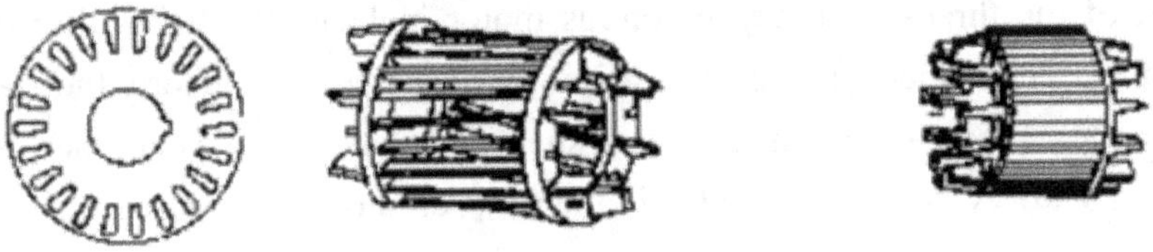

(a) Electronic core punch　(b) Rotor copper bar structure　(c) Schematic diagram for cast aluminum rotor

Fig. 2-3　Rotor structure of cage asynchronous motor

3) Rotation principle of three-phase asynchronous motor

(1) Demonstration experiment.

In order to explain how the rotor of three-phase asynchronous motor rotates, a demonstration experiment is done first.

As shown in Figure 2-4, there is a horseshoe magnet with a handle and a conductive squirrel cage rotor which can rotate freely in the middle of the magnetic pole. There is no mechanical connection between the rotor and the magnetic pole. When the handle is shaken to rotate the horseshoe magnet, the squirrel cage rotor will be seen to rotate with the magnet. The faster the handle rocks, the faster the rotor will be. If the rotation direction of the magnet is changed, the rotation direction of the squirrel cage rotor also changes.

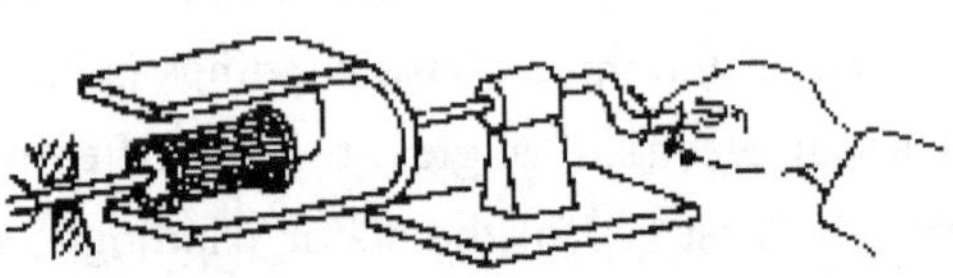

Fig. 2-4　Rotor rotation

Shown by the above experiment, having a rotating magnetic field is the first condition for rotor rotation.

(2) Generation of rotating magnetic field of stator.

In the actual cage asynchronous motor, the rotating magnetic field is generated by three-phase alternating current in the stator winding. If in the stator core of a three-phase asynchronous motor there are three relatively symmetrical windings U_1-U_2, V_1-V_2, and W_1-W_2, in a star-shaped (Y) connection, the current passes through the three-phase symmetrical windings when connected to a three-phase symmetrical power supply.

The waveform of three-phase symmetric current is shown in Figure 2-5. When the AC is positive half cycle, the current flows from the first end of the winding and flows out from the other end. In the negative half cycle, the current comes from the final end of the winding and outflows the first end. The inflow is represented by symbols "⊗", while the outflow is represented by symbols "⊙". When the alternating current flows through the three-phase windings, each phase winding will generate a magnetic field that changes in a sinusoidal pattern, and the resultant magnetic field of the three-phase windings will change directions over time to form a rotating magnetic field.

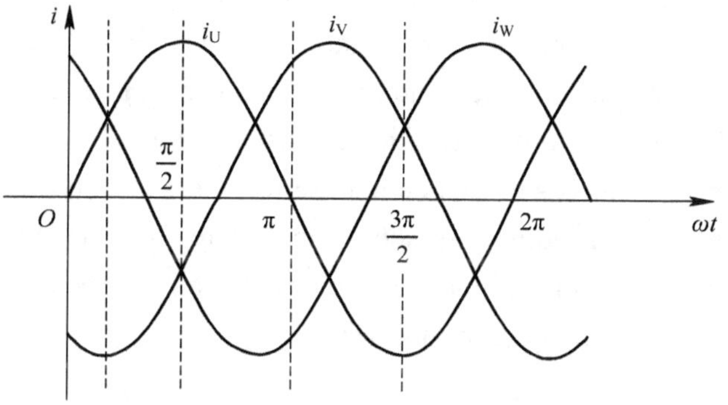

Fig. 2-5 Waveform of three-phase symmetrical current

When $\omega t = 0$, U phase winding current $i_U = 0$; when V phase winding current i_V is negative half cycle, according to the provisions, the current inflows from the final end V_2 and outflows from the first end V_1; when the current of the W phase winding i_W is positive half cycle, The current flows from the first end of the winding W_1 to the second end W_2. According to the right hand screw rule, it can be determined that the N pole of the resultant magnetic field of three-phrase current is directly above and its S pole is directly below.

When $\omega t = 90°$, i_U is positive half cycle; the current flows from the first end of the winding U_1 and flows out from the second end U_2. V phase and W phase currents i_V and i_W are negative half cycle. The currents flow from the winding ends V_2 and W_2 respectively, and flow out from the ends V_1 and W_1. It can be seen that the axis of the resultant magnetic field rotates 90° clockwise. At the moment, the N pole of the magnetic field is on the right, and the S pole is on the left.

When ωt=180°, U phase winding current i_U=0; when V phase winding current i_V is positive half cycle, according to the provisions, the current inflows from the first end V_1 and outflows from the second end V_2; when the current of the W phase winding is negative half cycle, the current flows in from the second end of the winding W_2 and out from the first end W_1. It can be determined that the N pole of the resultant magnetic field of the three-phase current is directly below and the S pole is directly above. The axis of the three-phase current's resultant magnetic field rotates another 90° clockwise.

It can be seen from the above analysis that when the stator windings of the asynchronous motor are connected with symmetric three-phase AC respectively, a rotating magnetic field with time duration can be generated in the stator space.

As shown in Figure 2-5, there are only two coils in each phase stator winding, and the space angle difference between the first ends of the three-phase windings is 120°. The resultant magnetic field has two poles, also known as a pair of poles. For a pair of magnetic poles, the magnetic field rotates once by a cycle in space when the three-phase alternating current changes once by a cycle. When the frequency of the AC current is 2 Hz, the magnetic field speed is 2 r/s; when the AC current frequency is 3 Hz, the magnetic field speed is 3 r/s; similarly, when the frequency of AC current is f Hz, the speed of magnetic field is f r/s. In general, the rotation speed of the rotating magnetic field is converted into revolutions per minute. The rotation speed (r/min) of the magnetic field for such a pair of magnetic poles is: $n_1 = 60f$.

If each phase winding consists of two coils in series, there is a space angle of 60° between the first ends of each phase winding. As shown in Figure 2-6, the magnetic field has four magnetic poles, namely two pairs of magnetic poles. It can be seen that the change in the alternating current by a cycle only goes around 180°.

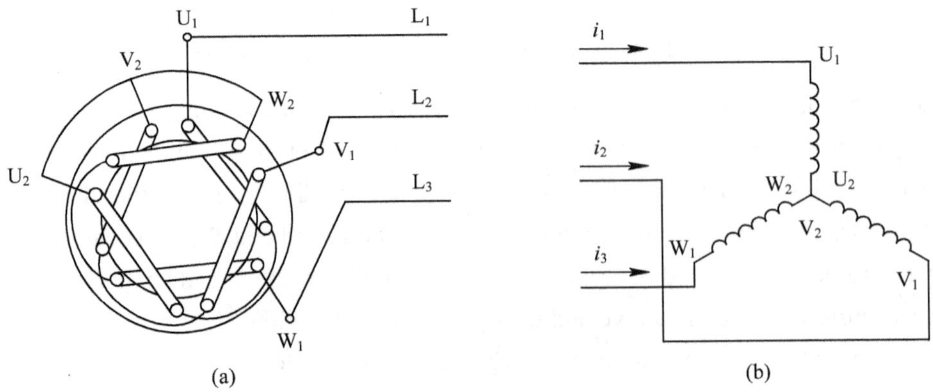

Fig. 2-6　Schematic diagram for winding

Similarly, if the rotating magnetic field has an arbitrary pair of magnetic poles, the alternating current changes by a cycle, and the rotating magnetic field can only rotate by $1/p$ cycle in space. The letters represent the number of the pair of magnetic poles of the rotating magnetic field. Therefore, the relationship between the rotational speed of the rotating magnetic field and the alternating current frequency and the number of the pair of magnetic poles is as follows:

$$n_1 = \frac{60f}{p}$$

where n_1—— the speed of the rotating magnetic field (also called the synchronous speed), with r/min as its unit;

f—— the frequency of the three-phase AC power supply, with Hz as its unit;

p—— the number of the pair of poles in a rotating magnetic field.

The rotation speed of the magnetic field decreases with the increase of the number of the pair of magnetic poles. Table 2-2 shows the rotation speed of the magnetic field corresponding to the number of the pair of magnetic poles if the power frequency is 50 Hz.

Table 2-2 Relationship between the number of the pair of magnetic poles and the rotation speed of the magnetic field when the power frequency is 50 Hz

The magnetic poles pair p	1	2	3	4	5	6
Rotating field speed n_1/(r/min)	3000	1500	1000	750	600	500

(3) Effect of rotating magnetic field on rotor.

The rotating magnetic field generated in the stator will cut the copper bar of the rotor. At this time, the magnetic field can be regarded as immovable and the rotor moves relative to the magnetic field. Assuming that the rotating magnetic field rotates clockwise, the rotor can be regarded as rotating counterclockwise relative to the magnetic field, as shown in Figure 2-7. Induced electromotive force and induced current are generated in the copper bar of the rotor, and their direction can be determined by the right-hand rule. In the upper half of the rotor, the direction of the induced current points to the reader, while in the lower half of the rotor, the direction of the induced current deviates from the reader.

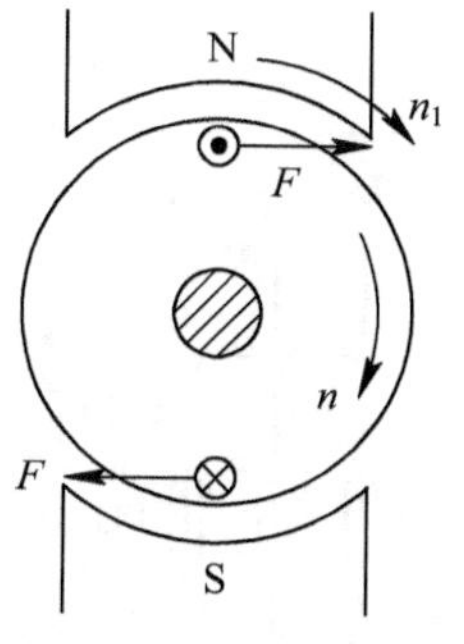

Fig. 2-7 Effect of rotating magnetic field on rotor

The copper bars loaded with induced current in the rotor interact with the rotating magnetic field to produce electromagnetic force. According to the left-hand rule, the force on the top copper bar of the rotor is on the right, and the force on the bottom copper bar is on the left.

These two forces are equal in magnitude and opposite in direction, forming electromagnetic torque, so the rotor rotates with the rotating magnetic field, which is the rotating principle of three-phase cage asynchronous motor. Rotor speed n must be less than synchronous speed n_1. If there is no relative motion between the rotor and the rotating magnetic field, that is, the copper bar on the rotor cannot cut the magnetic induction line, no induction electromotive force and induction current will be generated. And no electromagnetic torque will be formed, so the rotor cannot run at synchronous speed. In fact, the rotor speed is always less than the synchronous speed, that is, $n<n_1$. That is to say, the speed of the rotor is not synchronous with the speed of the rotating magnetic field, but asynchronous, which is the origin of the name of the asynchronous motor.

In normal operation, the rotor speed n is called the rated speed of the three-phase asynchronous motor. For example, there is a three-phase asynchronous motor with a pair of magnetic poles. Its synchronous speed n_1 is 3000 r/min and the rated speed n_e is 2906 r/min in normal operation.

The rotating direction of the rotor is the same as that of the rotating magnetic field. If the clockwise rotation is called the forward rotation of the motor, then the counter-clockwise rotation is called the reverse rotation of the motor. The direction of the rotating magnetic field is related to the phase sequence of the three-phase AC current in the stator winding. If the three-phase power supply is connected to any two relative positions of the three wires at the end of the stator winding, the rotating magnetic field will be reversed, and the motor will change in direction of rotation.

4) Use of three-phase asynchronous motor

(1) Start-up.

After the motor is connected to the power supply, the process of the rotor speed from zero to stable speed is called start-up. If the voltage applied to the stator winding of the motor at start-up is the rated voltage of the motor at work, the start-up is called full-voltage start-up, as shown in Figure 2-8.

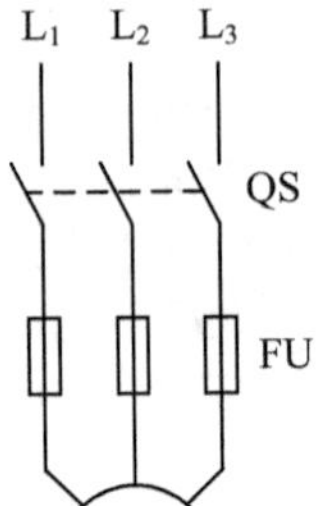

Fig. 2-8 Full-voltage start-up

At the beginning, the rotor has not started yet, but the rotating magnetic field has been generated. The magnetic field cuts the copper bar of the rotor at the maximum relative speed, which produces a large inductive current in the copper bar. Similar to the principle of transformer, the stator winding is equivalent to the primary winding of transformer and the copper bar of the rotor is equivalent to the secondary winding of transformer. Therefore, at the moment of motor start-up, a large start-up current will appear in the stator winding. Generally, the start-up current at full-voltage start-up is 4-7 times the rated current. When the motor is started infrequently, the starting time is very short (only 1-3 s). Although the current is very large, there is little effect on the motor. If the motor starts frequently, because of the accumulation of heat, the motor can be overheated, which can easily cause aging of insulation materials and shorten the service life of the motor. In addition, if the starting current of the motor is too large, the voltage drop of power supply line will increase in a short time, which will make the voltage at both ends of load drop in a short time. If it is a three-phase motor, the speed and the torque are reduced because of the voltage drop, so that the load cannot be driven, thus resulting in the phenomenon of stalling.

In practice, it is necessary to avoid the frequent starting of the motor as far as possible. For example, when turning, the friction clutch or electromagnetic clutch is used to separate the spindle from the motor shaft, thus reducing the starting and stopping of the motor, and avoiding the effect of excessive starting current on the service life of the motor.

Generally speaking, if the rated power of cage asynchronous motor is less than 7.5 kW, or the rated power is more than 7.5 kW and less than 20% the power supply capacity, the motor can be started by full-voltage.

If the line does not allow the motor to start at full-voltage, the step-down starting mode is adopted to limit the starting current. Step-down starting is the use of starting equipment to reduce the voltage appropriately, and then to apply it to the motor stator winding for start-up. And then the voltage is restored to the rated value after starting the motor. Step-down start-up is suitable for start-up under no load or light load.

There are four common methods for step-down starting: step-down starting of series resistor (or reactor) in stator winding, step-down starting of autotransformer, step-down starting of delta extension, and step-down starting of star (Y) triangle (△) transformation.

(2) Inversion.

Motor reversal is often required in production, as shown in Figure 2-9. When the switch closes upwards, the motor turns forward. When the switch closes downwards, the U and V phases of the three-phase power supply connected to the motor are exchanged to change the phase sequence of the three-phase AC current in the stator winding. Therefore, the reversal of the motor can be realized by changing the rotating magnetic field.

(3) Braking.

Braking means that after the motor is cut off from the power supply, it is given a torque in the opposite direction of rotation, which makes it slow down or stop quickly. For example, the crane hook needs to immediately decelerate or stop to achieve accurate positioning, and universal milling machine spindle quickly stops, etc., where braking is needed.

There are two braking methods: mechanical braking and electric braking.

Mechanical braking is a method of using mechanical devices to make the motor stop quickly after cutting off the power supply. Electromagnetic locking brake is the commonly used method as shown in Figure 2-10.

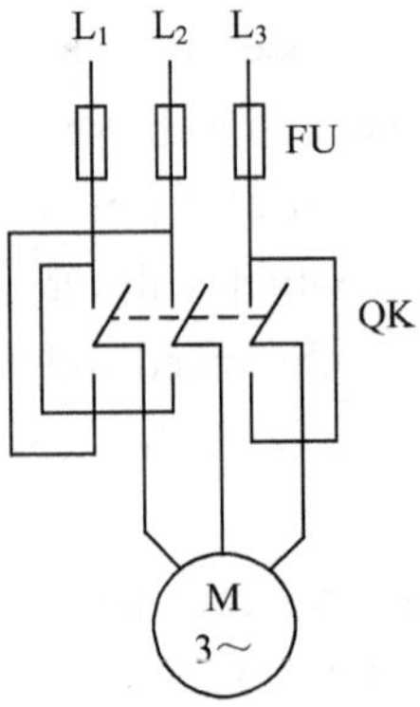

Fig. 2-9 Motor reversal

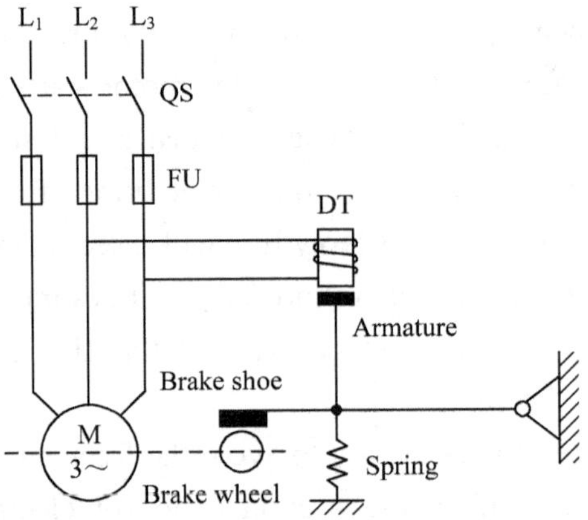

Fig. 2-10 Electromagnetic locking brake

The working principle of the electromagnetic locking brake is as follows: when the power supply is switched on, the coil of the electromagnetic locking brake is electrified and the armature is attracted, which overcomes the tension of the spring and forces the lever to move upward, so that the brake shoe and the brake wheel are separated. At this time, the motor starts and runs normally. Once the power of the motor is cut off, the coil of the electromagnetic locking brake loses power at the same time. Then the armature is released, under the action of spring tension, the brake shoe holds the brake wheel tightly, and the motor is quickly stopped by braking.

Electromagnetic locking brake method is widely used in lifting machinery. This braking method cannot only locate accurately, but also avoid accidents when the motor suddenly loses power.

The common methods of electric braking are reverse braking and energy dissipation braking.

Reverse braking is achieved by changing the power supply phase sequence of the input motor, so that the stator winding produces a reverse rotating magnetic field, and so that the rotor is subjected to the torque opposite to the original rotating direction and stops quickly. When reverse braking is adopted, it must be noted that when the speed of the rotor is close to zero, the power supply should be cut off in time, or otherwise the motor will run in reverse.

Energy dissipation braking is achieved when the motor running out of the power supply, DC current is immediately applied to its stator winding, which can make the motor braking. This method is stable in braking and accurate in positioning.

5) Nameplate data of three-phase cage asynchronous motor

At present, Y series three-phase asynchronous motors have been widely used in China. Take Y132M2-4 as an example to introduce the nameplate data.

(1) Type: Y series motor model consists of four parts. The first Chinese Pinyin letter Y represents the asynchronous motor; the second numbers part represents the center height (the height of axis center to installation platform surface); the third English letter part represents the seat length code (S means short seat, M means medium seat, and L means long seat); the number after the English letter is the core length code (1 means short core, and 2 means long core), and the number after the dash is the pole number of the motor; the fourth part is the special environmental code, the lack of the part meaning that the motor is suitable only for ordinary environment, W for outdoor environment and F for chemical anti-corrosion environment.

(2) Power: the power marked on the nameplate is the mechanical power output from the motor shaft under rated operating conditions, which is also known as capacity and often expressed by P_N or P_2 and measured in watts (W) or kilowatts (kW).

(3) Rated frequency: it refers to the electric frequency of the motor in rated operation, and the power frequency is 50 Hz as specified in China.

(4) Rated voltage: it refers to the value of the line voltage applied to the stator winding in rated operation of the motor, measured in volts (V).

(5) Rated current: it refers to the current value of stator winding of motor in rated operation, measured in amperes (A).

(6) Rated speed: it refers to the speed of the motor in rated operation, measured in r/min.

(7) Working methods: also known as quota, it refers to the operating state of the motor, indicated by continuous, short-term, and intermittent. "Continuous" refers to the continuous use of the motor for a long period of time in rated operation, represented by S_1. "Short-term" means that the motor runs for a short period of time within a specified time, represented by S_2. "Intermittent" means that the motor is operated in an intermittent mode, represented by S_3.

(8) Wiring: there are two ways to connect the specified sub-windings, star connection and triangle connection. Connect correctly according to nameplate mark. The junction box of cage three-phase asynchronous motor has six lead wires marked with U_1, V_1, W_1, U_2, V_2 and W_2. U_1 and U_2 are the two ends of the first phase winding, V_1 and V_2 are the two ends of the second phase winding, and W_1 and W_2 are the two ends of the third phase winding.

(9) Insulation grade: it refers to the heat-resistant grade of insulating materials, and is usually divided into seven grades.

In addition, the main technical data of three-phase asynchronous motor include power factor, efficiency, starting current, starting torque and maximum torque, but they are not marked on the nameplate but can be found in the product catalogue.

6) Key points for using three-phase asynchronous motor

Three-phase asynchronous motor is a kind of more durable motor, but as mechatronics equipment, it often goes out of order because of the frequency of use, working environment, and maintenance level. The user should be familiar with the key points for its use, to ensure the safe and normal work of the equipment.

Too frequent start and stop, or forward and reverse rotation will affect the service life and lead to overload. The abnormal sound inside the motor is the manifestation of faults. If it is not handled in time, it will lead to a damage of the motor. Mechanical faults, such as motor phase-absent operation, continuous overload and so on are accompanied by abnormal sound. Corresponding measures should be taken in time.When the motor emits a strange smell, it should be checked in time to prevent a larger fault.

2. Single-phase asynchronous motor

Single-phase asynchronous motors are widely used in industrial equipment and daily life, for example, fans, washing machines, dehydrators, refrigerators, household air conditioners, etc. They are powered by single-phase asynchronous motors. The power of single-phase asynchronous motors is small, ranging from a few watts to a few hundred watts.

The stator windings of single-phase asynchronous motor are single-phase, and the rotor is usually cage type. As shown in Figure 2-11, When the alternating current passes through the stator winding, the motor generates an alternating flux, which is sometimes vertically upward or vertically downward, and which is a pulsating magnetic field. In other words, the magnetic field generated by the winding of the single-phase stator is not a rotating magnetic field, so the rotor cannot rotate by itself. The motor has two stator windings, one is the working winding A_1A_2(also known as the main winding or the operating winding), and the other is the starting winding B_1B_2 (also known as the secondary winding). A_1A_2 and B_1B_2 differ by 90° in space and an appropriate capacitor is connected in the starting winding.

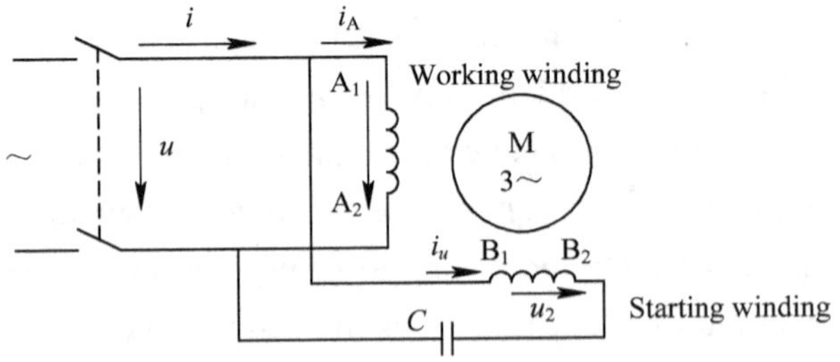

Fig. 2-11　Principle of single-phase asynchronous motor

When the motor is connected to the power supply, the windings are connected to the alternating current. The current i_A in the winding A_1A_2 lags behind the power supply voltage, while due to the capacitor connected in series with the winding B_1B_2, the current i_B leads the power supply voltage. If the capacitance of the capacitor is properly selected, a phase difference of 90° can occur between i_A and i_B, as shown in Figure 2-12(a). At this point, the stator windings of the motor generate a rotating magnetic field, as shown in Figure 2-12(b). The cage rotor rotates with the rotating magnetic field under the action of the rotating magnetic field.

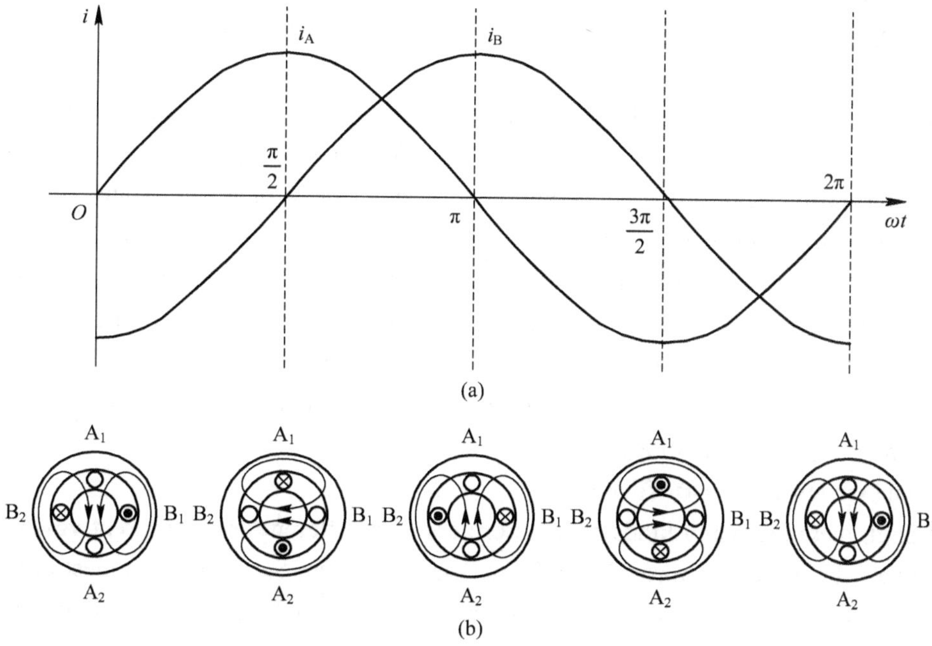

Fig. 2-12 Waveform diagrams for windings

In practice, please note that the capacitor is a fault-prone component. In replacing it, capacitors of the same capacity as the original and with a voltage rating no lower than that of the original should be selected.

3. Transformer

Transformers as well as AC motors are relatively durable appliances, which are in a special position in the specific application environment. The former usually supplies power to all devices, while the latter is usually the final execution device. Of special status and special role, they are often taken special care of. Problems with a transformer can affect the production of a factory and the supply of electricity to a community. Users should understand their characteristics and give them proper maintenance, so that they are kept in a normal working state.

Transformer is a kind of static electrical equipment which can change AC voltage and keep AC frequency unchanged.

Transformer is important electrical equipment in the process of power transmission and transformation. Transformers are usually used to increase the terminal voltage of generators when power is supplied. For the transmission of a certain power, the higher the voltage, the smaller the current, and the smaller the power loss over the transmission wire. With the small current, the transmission line with small cross-section can be selected, which can save a lot of metal materials. When using electricity, transformers are used to reduce the high voltage of the transmission line, so as to ensure personal safety and reduce the consumption of electrical insulation materials.

Normally, the voltage of an ultra-high voltage transmission line can be up to 500 kV (i.e., 500,000 V). However, in industrial and agricultural production and daily life we need a variety of different levels of AC voltage. For example, the rated voltage of the widely used three-phase asynchronous motor is 380 V or 220 V, the general lighting voltage is 220 V, and the rated voltage of the local lighting of the machine tool is 36 V, 24 V or lower. Many devices often require a variety of voltage supply. Therefore, in the actual work, we use a specified transformer to satisfy a different need. Apart from changing the alternating voltage, transformers can also change the alternating current (such as current transformers), transform the impedance (such as the input and output transformers in electronic circuits) and change the phase. Therefore, the transformer is the indispensable electrical equipment for transmission and distribution, electrical measurement, electronic technology and other aspects.

1) Basic structure of transformer

Although there are many kinds of transformers with different uses, different voltage levels and capacities, the basic structure of transformers is roughly the same. The simplest transformer consists of a closed soft magnet core and two mutually insulated windings, as shown in Figure 2-13.

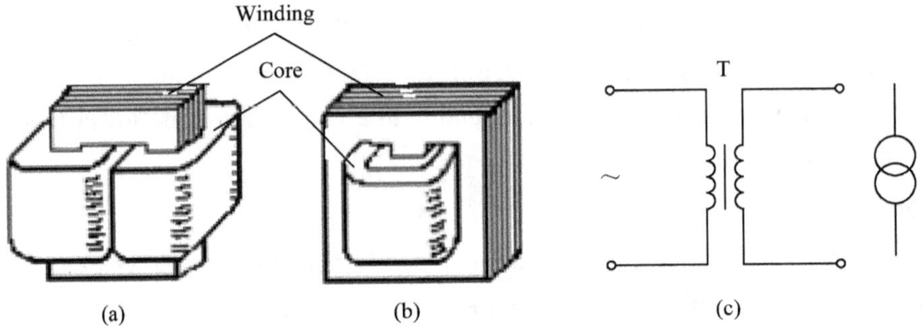

Fig. 2-13　Transformer mechanism diagram

Winding, also known as coil, is the circuit part of a transformer. The winding connected with AC power supply is called primary winding, and the winding connected with load is called secondary winding, as shown in Figure 2-14.

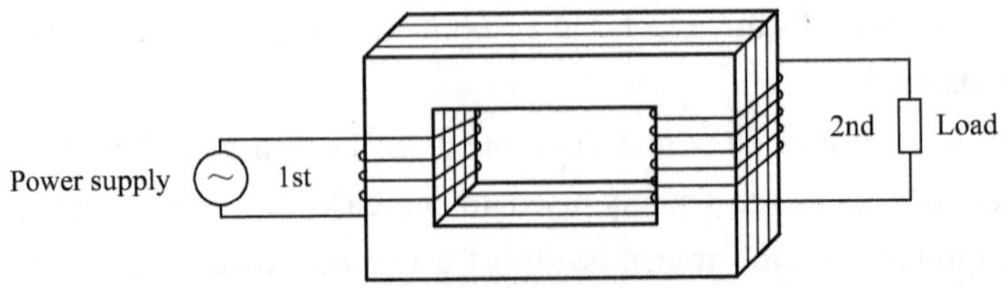

Fig. 2-14　Secondary winding diagram

Core is the magnetic circuit part of a transformer, which is a stack of silicon steel sheets of the thickness of 0.35-0.5 mm. According to the core structure and winding configuration of a transformer, there are two types of transformer, core type and shell type. As shown in Figure 2-15(a), the single-phase core transformer uses a mouth-shaped core. The primary and secondary windings are respectively wound on the core. The single-phase shell transformer is commonly used with hill-shaped and sun-shaped cores, as shown in Figure 2-15(b).

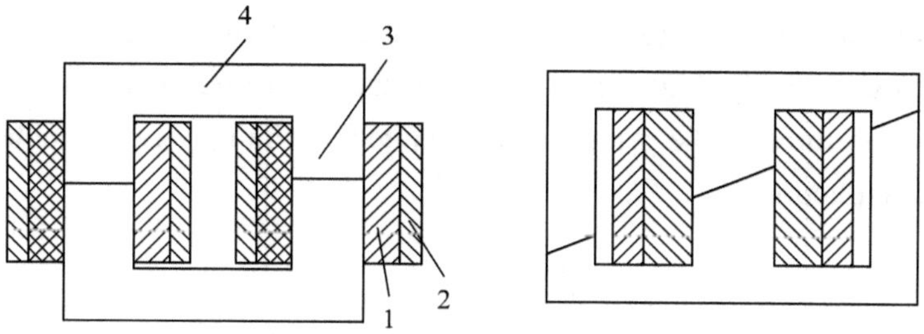

(a) Core transformer (b) Shell transformer

1— Primary winding; 2—Secondary winding; 3—Iron core; 4—Magnetic yoke

Fig. 2-15 Single-phase core type transformer

2) Working principle of transformer

Figure 2-16 shows the schematic diagram for the working principle of single-phase transformer. For convenience, please note these rules: all quantities related to the primary winding are marked with "1" in the lower right corner of their symbols, while those related to the secondary winding are marked with "2" in the lower right corner of their symbols. For example, the primary and secondary voltage, current, turn number and electromotive force are represented by U_1, U_2, I_1, I_2, N_1, N_2, E_1 and E_2 respectively.

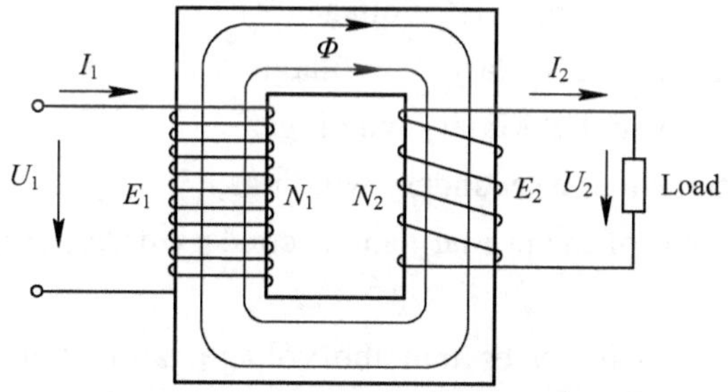

Fig. 2-16 Working principle diagram for single-phase transformer

When the primary part of a transformer is connected to the AC power supply, there is an alternating current flowing through the primary windings, and then an alternating flux is generated in the core, which is called the main flux. The main flux varies with the frequency of the power supply, and is concentrated in the core. A very small part of it is closed outside the primary windings, known as the flux leakage, which is generally very small and can be neglected. According to the law of electromagnetic induction, the primary and secondary windings will produce induced electromotive force. If the secondary is connected with a load to form a closed loop, an induced current is generated. The transformer transmits the power supply energy to the load through the magnetic coupling of the primary and secondary windings.

(1) Principle of voltage transformation by transformer.

Set the number of turns of the primary and secondary as N_1 and N_2 respectively, and neglect the leakage flux and the influence of the primary and secondary DC resistance. Because the primary and secondary windings are both affected by the main magnetic flux, the frequency of the induced electromotive potential E_1 and E_2 generated in the two windings are the same as the frequency of the power supply. If the change rate of the main flux over time is $\Delta\Phi/\Delta t$, then the induced electromotive force of the primary and secondary windings can be obtained by the law of electromagnetic induction:

$$E_1 = \left| N_1 \frac{\Delta\Phi}{\Delta t} \right| \tag{2-1}$$

$$E_2 = \left| N_2 \frac{\Delta\Phi}{\Delta t} \right| \tag{2-2}$$

The terminal voltage of the primary and secondary of the transformer is approximately equal to the induced electromotive force in numerical value. Therefore, after taking into account the above two expressions and not considering the phase relationship, and only considering the magnitude, the relationship between effective values of the primary and secondary voltages can be obtained as follows:

$$\frac{U_1}{U_2} = \frac{N_1}{N_2} = n \tag{2-3}$$

where U_1——effective value of primary AC voltage (V);

U_2—— effective value of secondary AC voltage (V);

N_1——number of turns in the primary windings;

N_2——number of turns in the secondary windings;

n—— the voltage ratio of the primary and secondary order, or the ratio of the number of turns.

The above formula (2-3) indicates that the voltage ratio of the primary and secondary windings of a transformer is equal to that of their number of turns. When $n > 1$, and $N_1 > N_2$, then $U_1 > U_2$. This transformer is a step-down transformer. When $n < 1$, and $N_1 < N_2$, then $U_1 < U_2$. This transformer is a booster transformer. Therefore, as long as a different ratio of the number of turns of the primary and secondary windings is selected, the requirements for boost or step-down can be met.

Example 1: The primary windings of a transformer are connected to a high-voltage transmission line of 10 kV. The secondary windings are required to output a voltage of 400 V. If the number of turns of the primary windings is 800, find out the voltage ratio of the transformer and the number of turns of the secondary winding N_2.

By the formula (2-3), the voltage ratio is:

$$n = \frac{U_1}{U_2} = \frac{10000}{400} = 25$$

and then

$$N_2 = \frac{N_1}{n} = \frac{800}{25} = 32$$

(2) Principle of current transformation by transformer.

Any kind of transformer only plays the role of energy transfer in the process of voltage transformation. No matter whether the voltage after transformation is increased or decreased, the electric energy will not increase or decrease. According to the law of conservation of energy, when the loss is neglected, the output power P_2 of the transformer should be equal to the power P_1 obtained by the transformer from the power supply, that is,

$$P_1 = P_2$$

or

$$U_1 I_1 = U_2 I_2$$

so

$$\frac{I_1}{I_2} = \frac{U_2}{U_1} = \frac{N_2}{N_1} = \frac{1}{n} \tag{2-4}$$

The formula (2-4) is the formula for current transformation by transformer. This formula indicates that when the transformer is working, the current value of the primary and secondary windings is inversely proportional to the voltage or the number of the turns of the primary and secondary or the inverse of the transformer voltage ratio. In fact, the transformer changes the current as well as the voltage. Current transformers are made based on the principle.

Example 2: A transformer is connected to an AC circuit of 3300 V. If the voltage is lowered to 220 V, the number of turns of the primary windings is known to be 2100, and if the load of the secondary windings is 10, find out the number of turns of the secondary windings and the current of the primary and secondary windings.

According to the formula (2-3), the voltage ratio of the transformer is

$$n = \frac{U_1}{U_2} = \frac{3300}{220} = 15$$

The number of turns of the secondary windings is

$$N_2 = \frac{N_1}{n} = \frac{2100}{15} = 140 \text{ (turns)}$$

According to Ohm's law, the current of the secondary windings is:

$$I_2 = \frac{U_2}{R_2} = \frac{220}{10} = 22(\text{A})$$

From formula (2-4), it can be concluded that the current of the primary windings is:

$$I_1 = \frac{I_2}{n} = \frac{22}{15} \approx 1.47(\mathrm{A})$$

3) Several common transformers

(1) Single-phase lighting transformer.

A common single-phase lighting transformer, as shown in Figure 2-17, consists of an iron core and two mutually insulated coils, and is usually of a shell type. The rated primary voltage of this transformer is 220 V and 880 V, and the secondary voltage is mostly 36 V. When used in special dangerous situations, the secondary voltage is 24 V or 12 V. Some transformers have a secondary voltage of 6 V, which is used only for indicator light. The single-phase lighting transformer often provides safe voltage for local lighting lamps in a factory to ensure personal safety.

Fig. 2-17 Single-phase lighting transformer

(2) Three-phase transformer.

The three-phase transformer is widely used in industrial production. The so-called three-phase transformer is essentially composed of three single-phase transformers of the same capacity. Figure 2-18 is a schematic diagram for the three-phase transformer. The primary (high voltage) winding and the secondary (low voltage) winding of the same phase are wound on each core column.

In terms of the different conditions of the three-phase power supply and the load, the primary and secondary windings of the transformer can be connected in a Y shape or a $\triangle$ shape. Figure 2-18(a) shows the standard wiring diagram for a three-phase transformer. For large-capacity three-phase transformers, Y/$\triangle$ connection is used, that is, the high-voltage winding is Y-shaped connection, and the low-voltage winding is $\triangle$-shaped connection. This is because the phase voltage of the Y-shaped junction is $1/\sqrt{3}$ the line voltage, which is beneficial to the coil insulation. However, the low-voltage winding has a $\triangle$-shaped junction, which makes the conductor cross section smaller than that of the Y-shaped junction. For transformers of small capacity and requiring a neutral line, Y/Y_0 connection is often used, that is, the high-voltage winding is Y-shaped and the low-voltage winding is Y-shaped and has a neutral line. This connection allows the user to obtain both line voltage and phase voltage, especially suitable for loads mixed with power and lighting.

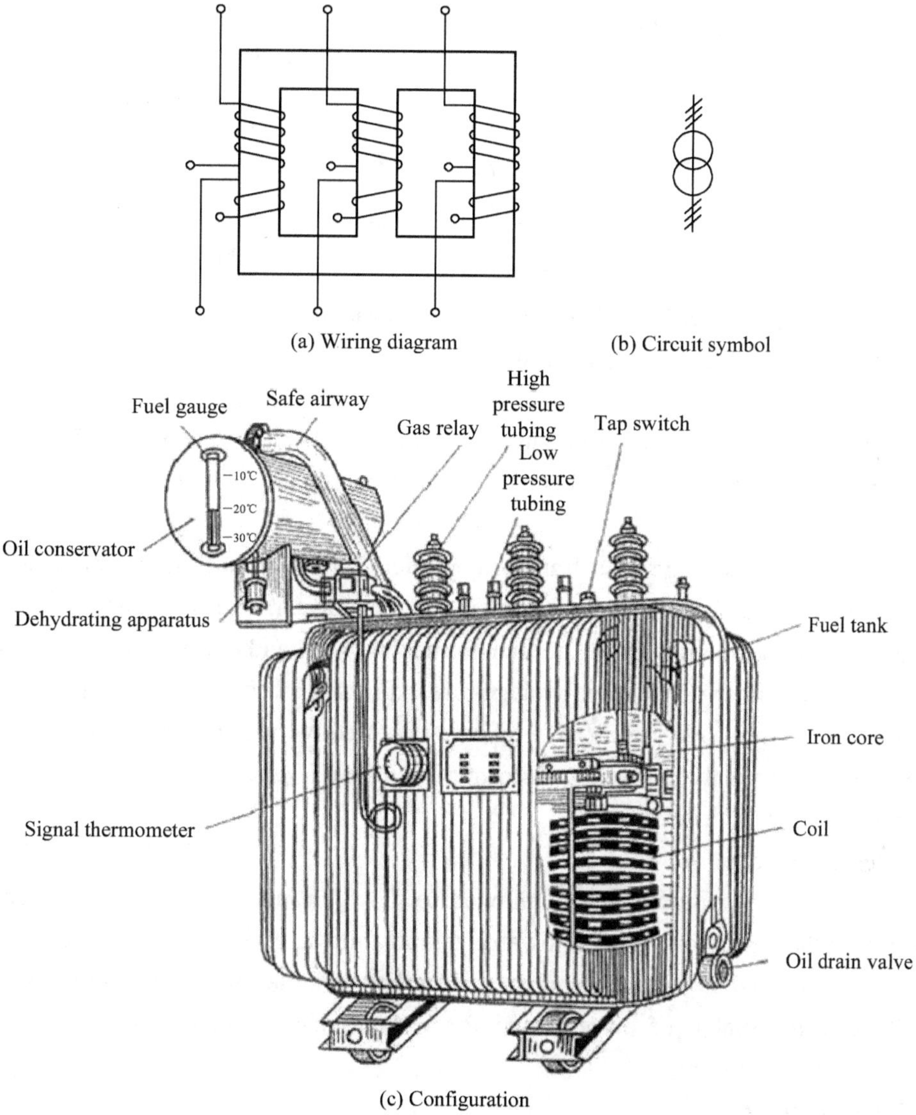

Fig. 2-18 Schematic diagram for three-phase transformer

(3) Auto-transformer.

Figure 2-19 is a schematic diagram for the auto-transformer. The auto-transformer has a ring core with the coil wound around the core, that is, there is only one winding. A part of the primary and secondary windings is shared, that is, a part of the high-voltage winding acts as a low-voltage winding. It is the same as the general transformer.The voltage ratio of the primary winding to the secondary winding is equal to the ratio of the number of turns of the primary to the secondary. The output point of the secondary voltage is a brush contact which can slide freely along the exposed surface of the coil. The desired output voltage can be obtained by changing the position of the contact. The auto-transformer is commonly used in the step-down starting equipment of laboratory and AC asynchronous motor. Its biggest feature is that the required voltage can be obtained through "voltage regulation".

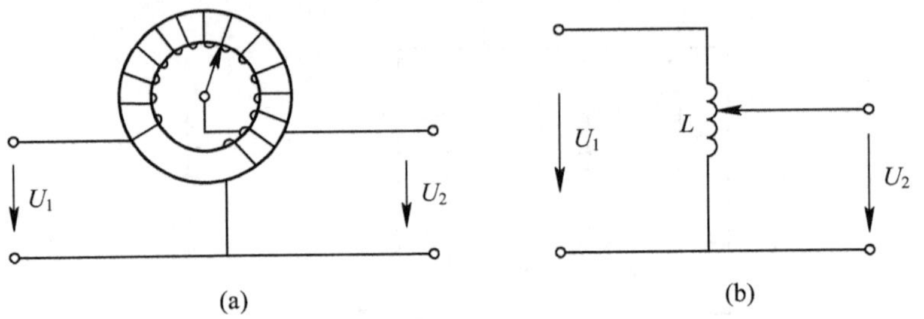

Fig. 2-19 Schematic diagram for auto-transformer

4) Main technical data of transformers

The specifications, types and main technical data of a transformer are marked on its nameplate, which serves as an important basis for its use. The main technical data of a transformer include rated voltage, rated current, rated capacity and temperature rise.

(1) Rated voltage.

The primary rated voltage of a transformer refers to the voltage value stipulated by the insulation strength of insulating materials used in the transformer. The secondary rated voltage is the voltage value across both ends of the transformer working without a load after applying the rated voltage to the primary. Two rated voltages are expressed as U_{1N} and U_{2N} respectively. For a single-phase transformer, U_{1N} and U_{2N} refer to the effective value of primary and secondary AC voltage, while for a three-phase transformer, U_{1N} and U_{2N} refer to the effective value of primary and secondary line voltage.

(2) Rated current.

It refers to the maximum current allowed to flow through the primary and secondary windings under the condition of an allowable temperature rise. The secondary current of the transformer is expressed as I_{1N} and I_{2N} respectively. For a single-phase transformer, I_{1N} and I_{2N} refer to the effective value of current, and for a three-phase transformer, they refer to the effective value of line current.

(3) Rated capacity.

It represents the maximum power allowed to be transferred when the transformer is working. The rated capacity of single-phase transformer is the product of secondary rated voltage and rated current; the rated capacity of three-phase transformer is also the product of secondary rated voltage and rated current (should be the sum of three-phase). The rated capacity is denoted by the letter S measured in volt-amperes (V • A).

(4) Temperature rise.

Temperature rise refers to during the rated working time of the transformer, an allowable rise in temperature which exceeds the ambient temperature. It depends on the heat resistance level of insulating materials of the transformer, as shown in Table 2-3.

Table 2-3 Heat resistance level of insulating materials

Insulation level	Y	A	E	B	F	H	C
Maximum operating temperature/℃	90	105	201	130	155	180	>180

5) Key points of transformer use

Transformers used in electrical applications are generally for high and low voltage conversion of electrical energy. A transformer often converts energy for a community, a factory, a workshop and a set of equipment. The natural environment, its quality, the change of its load and the condition of its maintenance may affect its work. Minor accidents may affect the operation of transformers and may scrap the equipment. Large transformer accidents are likely to cause explosion, thus endangering personal safety.

Large special transformers are maintained and supervised by special personnel. Generally, the relevant personnel should pay attention to the change of the external characteristics of transformers, and report to the relevant personnel in time.

Transformers working normally generally make some slight vibration sound, allow a certain temperature rise, and do not send out an odor. Once the vibration sound increases significantly or there are special phenomena such as strange smell and fire, it is necessary to report to the personnel in charge and keep away in time.

【Practice and Operation】

1. Task preparations

Materials and equipment needed for the experiment are prepared for inching and continuous rotation control circuit.

2. Task circuit

Connect according to Figure 2-20, and the experiment can be carried out only after the instructor checks it.

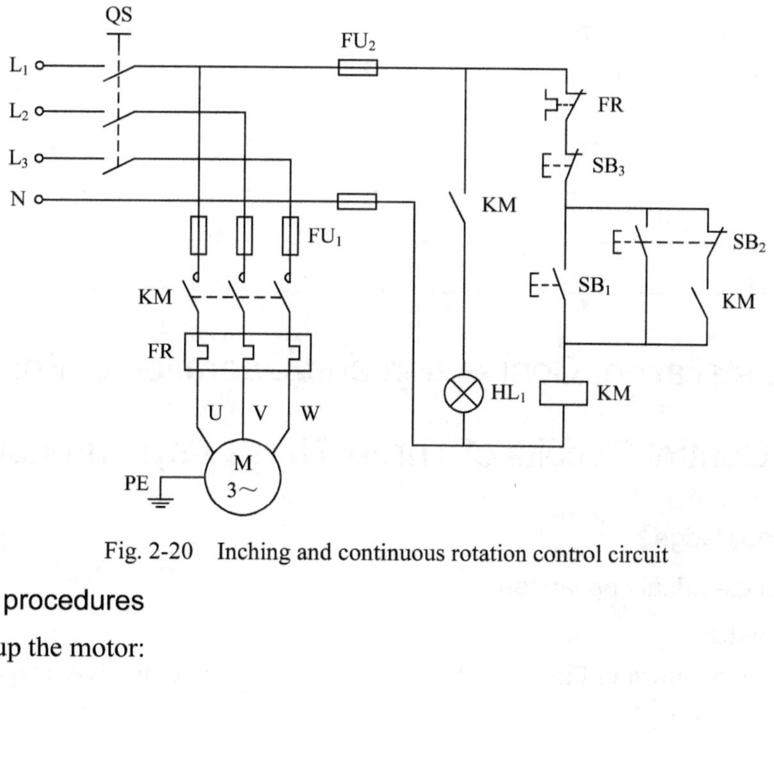

Fig. 2-20 Inching and continuous rotation control circuit

3. Task procedures

(1) Start up the motor:

Switch on QS → press SB_1 → KM is on →
- → KM closed self-locking → HL_1 is on
- → KM self-locking contact closed self-locking
- → KM main contact closed → Motor starts

(2) Operate the motor:

At the instant of pressing SB_2 → SB_2 normally connected contact is opened, and the self-locking circuit of KM is disconnected, resulting in the loss of power of the KM coil (due to the inertial, the motor continues torotate) → fully press SB_2 → KM gets electricity →
- → KM normally open contact closed → HL_1 is on
- → KM main contact closed, the motor is operated (KM normally open auxiliary contact is closed, but self-locking function is invalid)

(3) Stop the motor:

Press SB_3 → KM loses electricity → Contacts are reset →
- → HL_1 is off
- → The motor stops

4. Precautions

(1) Recheck the accuracy of wiring according to inching and continuous rotation control circuit diagram.

(2) When starting the DC motor, first turn on the excitation power switch, and then turn on the armature power switch. When stopping the motor, first turn off the armature power switch, and then turn off the excitation power switch.

5. Notes and summary

Task 2　Installation, Connection and Operation of Forward and Reverse Control Circuits of Three-Phase Asynchronous Motor

【Relevant Knowledge】

1. Manual electrical appliances

1) Knife switch

Knife switch is shown in Figure 2-21, which is generally used in low-voltage circuits with infrequent operation. It is used to turn on and off a power supply, and sometimes to control the direct start and stop of a small-capacity motor.

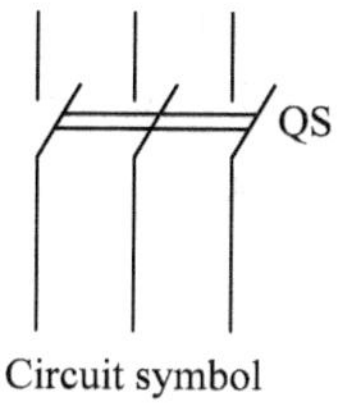

Fig. 2-21 Circuit symbol for knife switch

The knife switch is composed of a brake knife (moving contact), a static socket (static contact), a handle and an insulating base plate.

There are many kinds of knife switches. They fall into unipolar, bipolar and tripolar according to the number of poles (blades); there are flat and strip frame according to structure; there are direct handle operation, lever operation and electric operation according to operation mode; there are single and double throw according to conversion direction.

The knife switch is usually used in series with the fuse in order to cut off the circuit automatically when the fuse is fused in short circuit or overload.

The rated voltage of the knife switch is usually 250 V and 500 V, and the rated current is below 1500 A. Considering the large starting current of the motor, the rated current value of the knife switch should be chosen to be 3-5 times the rated current of the asynchronous motor.

2) Button

Buttons are often used to turn on and off control circuits. Their structure and circuit symbol are shown in Figure 2-22. The contacts on buttons are divided into normally open contacts and normally closed contacts. Because of the structural characteristics of buttons, buttons are used only for sending "on" and "off" signals.

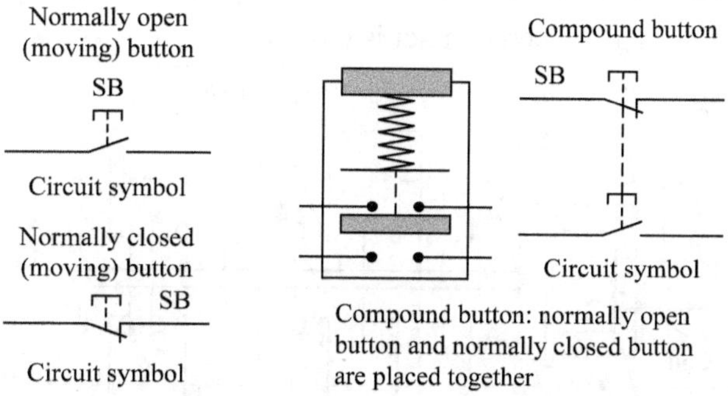

Fig. 2-22 Button structure diagram

2. Automatic electrical appliances

1) Fuse

The fuse is mainly used for short circuit or overload protection and is connected in series in the protected circuit. The circuit symbol for fuse is shown in Figure 2-23. When the line works normally, it acts as a channel like a wire; however, when the line is short-circuited or overloaded, the fuse breaks and protects other electrical equipment along the line. The structures of the fuse are tube type, magnetic insertion type, spiral type and so on. Its core part of the melt (fuse or ferrule) is made of fusible alloys of high resistivity, such as lead-tin alloys, or of conductors with small cross-sectional area.

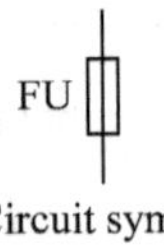

Fig. 2-23　Circuit symbol for fuse

Selection of the rated current I_F for the melt is done as follows:

(1) On non-impact current occasions (e.g., electric lamps and electric furnaces), $I_F \geqslant I_L$.

(2) The melt of a motor: the rated current of the melt is greater than or equal to the starting current of the motor ÷2.5; if the motor starts frequently, the rated current of the melt is greater than or equal to the starting current of the motor ÷ (1.6–2).

(3) The total melt for several motors: the rated current of the melt is equal to(1.5–2.5) × the rated current of the motor of the largest capacity + the sum of the rated currents of the other motors.

2) AC contactor

Contactor is an automatic switch, which is one of the main control devices in electric drive. It can be divided into DC and AC. AC contactor is often used to connect and disconnect the main circuit of a motor or other equipment. Figure 2-24 is the main structure diagram for AC contactor. Contactors are mainly composed of electromagnets and contacts, driven by the attraction of electromagnets. When the electromagnetic coil is electrified, the hill-shaped moving core (upper core) is attracted and the normally open contact is closed.

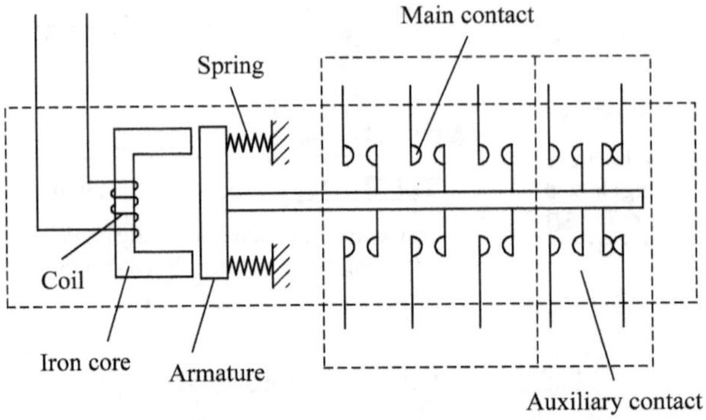

Fig. 2-24　Working principle of contactor

According to the different uses, the contacts of contactors are divided into main contacts and auxiliary contacts. The current passing through the auxiliary contacts is small and they are often connected to the control circuit of the motor. The main contacts allow a larger current to pass through and are often connected to the main circuit of the motor. For example, CJ10-20 AC contactor has three pairs main contacts and two pairs of auxiliary contacts (one pair of normally open contacts and one pair of normally closed contacts).

When the main contacts are disconnected, arc will occur during the disconnection, which will burn the contacts and prolong the disconnection time of the circuit. Therefore, arc extinguishing measures must be taken. Usually the contacts of AC contactors are made to be of bridge structure. There are two breakpoints to reduce the voltage applied to the breaking point when the contacts are disconnected, so that the arc can be easily extinguished. At the same time, insulating baffles are installed between phases to prevent short circuit. The arc extinguishing device is specially installed in contactors with larger current. The circuit symbol for the contactor is shown in Figure 2-25.

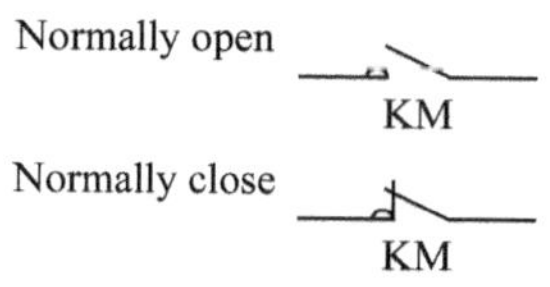

Fig. 2-25 Circuit symbol for contactor

When choosing a contactor, attention should be paid to its rated current, coil voltage and the number of contacts. The rated currents of main contacts of CJ10 series contactors are 5 A, 10 A, 20 A, 40 A, 75 A, 120 A and so on.

3) Intermediate relay

The structure of the intermediate relay is basically the same as that of the contactor, but it is small in size and has more contacts. It is usually used to transmit signals and control multiple circuits at the same time. It can also be used to control small-capacity motors or other actuators.

The commonly used intermediate relays are JZ7 series. The rated current of the contacts is 5 A. The voltage of the coil should be considered when using them.

4) Thermal relay

Thermal relay is used to protect a motor from the damage caused by long-term overload.

Thermal relay is operated by the thermal effect of current. Its working principle is shown in Figure 2-26. In the figure, the thermal element is a resistor wire of small resistance. Connected in the main circuit of the motor, the bimetal sheet is made of two metals of different linear expansion coefficients rolled by heat and pressure or bonded by cooling. The expansion coefficient of the lower metal is larger than that of the upper metal. When the current through the main circuit exceeds the allowable value, the bimetal sheet bends upward when heated, which causes the tripping. The buckle plate will break the normally closed contacts under the tension of the spring. Contacts are connected to the control circuit of the motor. When the control circuit is disconnected, the coil of the contactor is cut off, and thus the main circuit of the motor is disconnected.

Due to thermal inertia, the thermal relay cannot be used for protection against short circuit. In case of short circuit accidents, we require the circuit to be immediately disconnected, while the thermal relay cannot be immediately activated. However, this thermal inertia is in line with our requirements. For example, when the motor starts up or is temporarily overloaded, unnecessary stopping of the motor can be avoided because with the thermal inertia the thermal relay will not operate. To reset the thermal relay, press the reset button.

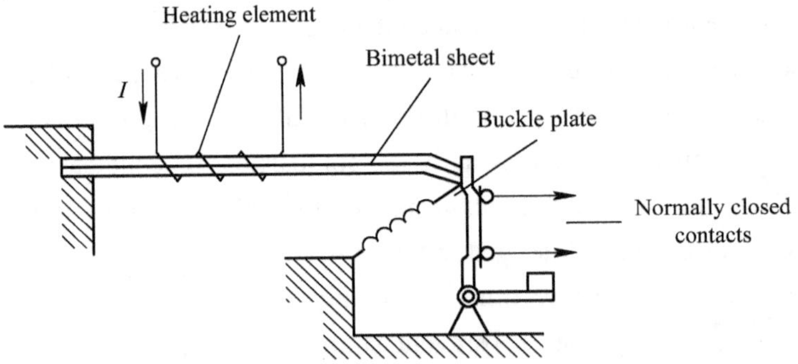

Fig. 2-26　Working principle of thermal relay

Commonly used thermal relays are JR0, JR10 and JR16 series. The main technical data of thermal relay are setting current. The setting current means that when the current passing through the heating element exceeds 20% of this value, the thermal relay should operate within 20 min. The setting current of JR0-40 type ranges from 0.6 A to 40 A and there are 9 specifications. When selecting the thermal relay, its setting current should be basically the same with the rated current of the motor.

5) Stroke switch

The structure of the stroke switch is similar to that of the button, but it is activated by mechanical shock. It is used as limit protection, stroke control, and automatic switching for circuits, as shown in Figure 2-27.

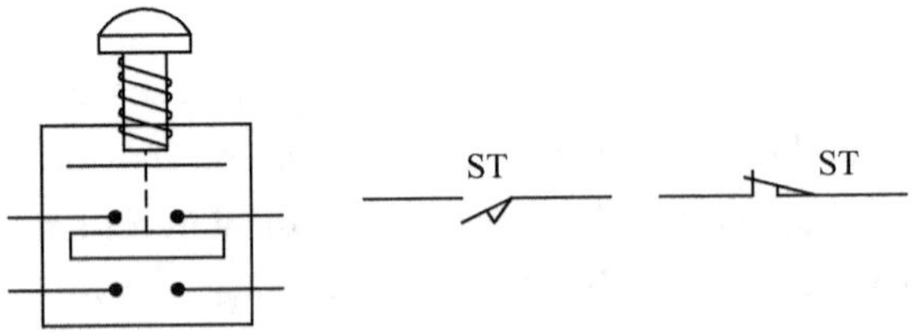

Fig. 2-27　Structural diagram of and circuit symbol for stroke switch

【Practice and Operation】

1. Task preparations

According to the experiment content of the forward and reverse control circuit of a contactor interlocking motor, prepare the materials and equipment required for the experiment.

2.Task circuit

According to the forward and reverse control circuit of a contactor interlocking motor, as shown in Figure 2-28, the wiring should be done and checked by the instructor before practicing the task procedures.

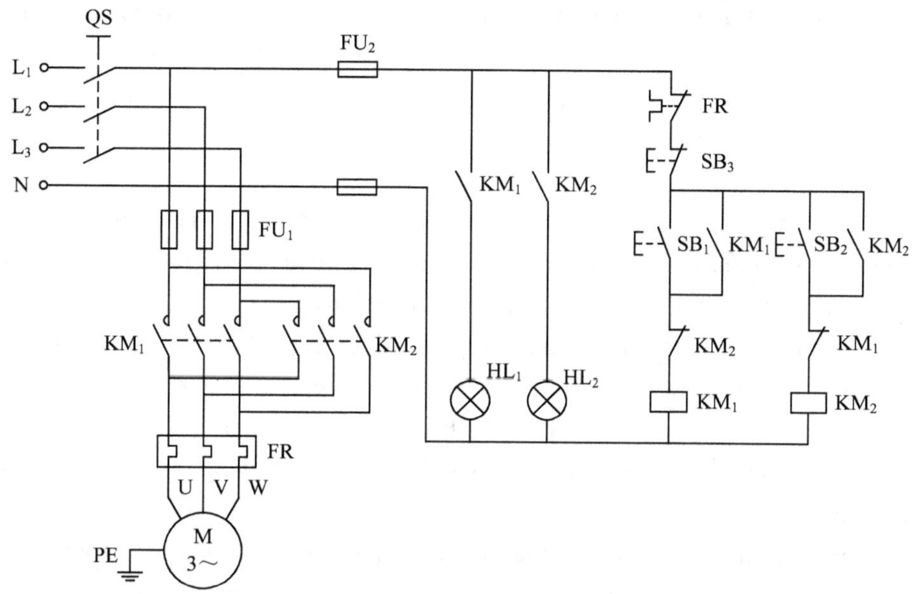

Fig. 2-28 Forward and reverse control circuit of contactor interlocks motor

3. Task procedures

(1) Forward rotation of a motor:

Switch on QS → Press SB_1 → KM_1 is powered on
- → KM_1 main contact is closed → The motor forward rotates
- → KM_1 normally opened self-locking contact closes self-locking → HL_1 is on

(2) Reverse rotation of a motor:

Press SB_2
- → KM_1 is powered off
  - → KM_1 main contact is disconnected → The motor loses electricity
  - → Each auxiliary contact is reset
- → KM_2 is powered on
  - → KM_2 normally closed contact is disconnected with KM_1 interlocking, KM_2 main contact is closed → The motor inversely rotates
  - → KM_2 normally opened self-locking contact closes self-locking → HL_2 is on

(3) Press SB_3→The motor stops.

4. Precautions

(1) The reasonable arrangement of the positions of electrical appliances and instruments before wiring is usually based on the principle of easy operation and easy observation of readings. The distance between the electrical appliances is appropriate, with neat and beautiful wiring and easy inspection as the criteria.

(2) After the connection of the positive and reverse control lines of the contactor

interlocking motor is completed, a comprehensive inspection should be carried out. If there is no mistake, please ask the instructor to check, and then switch on the power to start the test.

(3) In the connection training, we should grasp the general control rules, such as series connection before parallel connection, main circuit before control circuit, control contact, and then protection contact before control coil, etc.

(4) It is the most important step in the whole experiment process to observe various phenomena or record experimental data in the forward and reverse control circuit experiment of the contactor interlocking motor, which must be taken seriously.

(5) In the training of control circuit, we should operate the main command electrical apparatus purposefully, observe the action of the electrical apparatus, and further understand the working principle of the circuit. If abnormal phenomena occur, the power supply should be switched off immediately, inspection and analysis should be carried out, and experiments should be continued after troubleshooting.

(6) At the end of the training, the power supply should be disconnected, and the experimental results should be carefully checked to ensure that there are no omissions or other problems. After the instructor's examination and approval, the circuit can be dismantled, the experimental equipment, wires and tools can be cleaned up.

5. Notes and summary

____

____

____

____

Task 3 Installation, Connection and Operation of Y-△Start Control Circuit of Three-Phase Asynchronous Motor

【Relevant Knowledge】

1. Y-△ step-down start

The purpose of Y-△ starting is to carry out step-down starting, to prevent the excessive voltage at starting and then protect the motor, also known as Y-△ step-down starting. Step-down starting is to use the starting equipment to reduce the voltage properly and then apply it to the stator winding of the motor to get started. After the motor starts running, the voltage will return to the rated value for normal operation. Step-down starting can reduce the starting current, but the starting torque is also reduced (because the torque of the motor is proportional to the square of the voltage), so step-down starting is mostly used for no-load or light-load starting of a squirrel cage motor.

Step-down starting method: in normal operation, when the load has no strict requirements for the starting torque of the motor, and when the starting current of the motor is limited, for the squirrel cage asynchronous motor whose stator winding is connected into a triangle, Y-△ step-down starting can be used. If the stator winding of the motor is connected in a triangle way, its six leads are connected to the Y-△ starter. At starting, connect the stator winding to a Y, and change it to a △ when the motor speed is close to the rated speed.

2. Y-△ wiring of motor

1) Y connection (star connection)

Star connection is a method for the connection of three-phase AC power supply and three-phase electrical appliances. Connect the terminal ends of the three windings of the three-phase power supply together to form a common point, and lead out three end lines from the beginning. The three-phase alternating current is a system which is connected by three sinusoidal power supplies of the same frequency, the same amplitude and a phase difference of 120° in turn.

2) △ connection(triangle connection)

The triangle connection method is to connect the power supply or load of each phase in turn, and lead out each connected point as the three phase lines of three-phase electricity. Being like the shape of a triangle, this connection is called a triangle connection.

【Practice and Operation】

1. Task preparations

Prepare necessary materials and equipment according to the training contents of Y-△ start control circuit of three-phase asynchronous motor.

2. Task circuit

Connect the circuit according to Figure 2-29. After the instructor checks it, the experiment can be carried out.

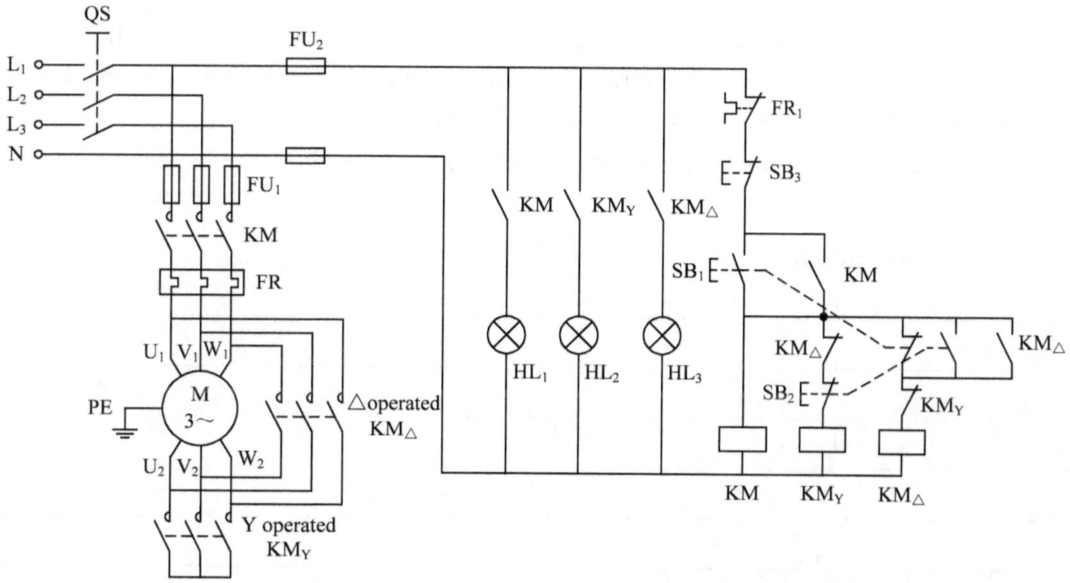

Fig. 2-29　Y-△ start control circuit of three-phase asynchronous motor

3. Task procedures

(1) Steps for Y connection starting are as follows:

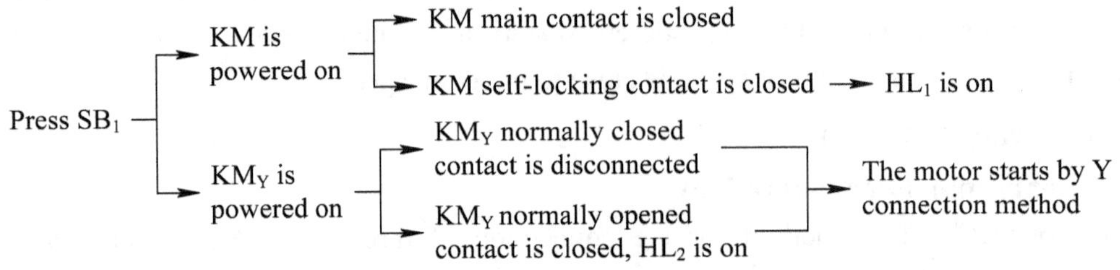

(2) When the speed of the motor rises to a certain value, the full-voltage operation of the motor by △ connection method is achieved. The steps are as follows:

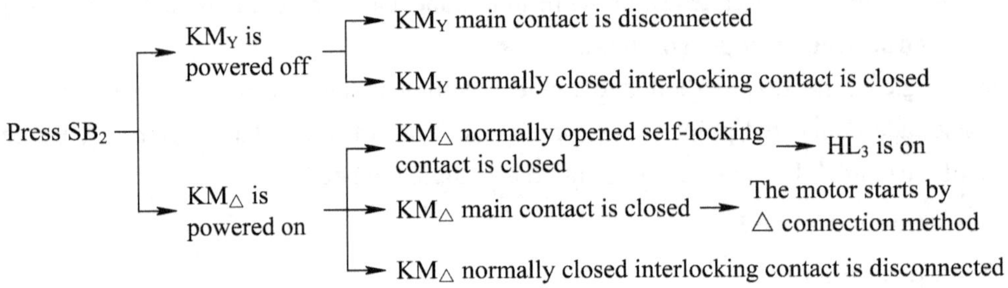

(3) Prcss SB_3, and thc motor stops.

4. Precautions

(1) Frequent start-up and stop, or forward and reverse, will affect the service life and lead to overload.

(2) The abnormal sound inside the motor is the manifestation of faults. If they are not handled in time, they will lead to the damage of the motor. Mechanical faults, such as motor phase-absent operation, continuous overload, bearing faults and so on, are accompanied by abnormal sound. Corresponding measures should be taken in time.

(3) When the motor emits a strange smell, it should be checked in time to prevent the fault from expanding.

5. Notes and summary

Task 4 Installation, Connection and Operation of Direct Switching Control Circuit of Two-Speed Three-Phase Asynchronous Motor

【Relevant Knowledge】

Two-speed three-phase asynchronous motor operates by asynchronous motor pole-changing speed regulation, which is achieved by changing the method for the connection of stator windings to change the number of pole pairs of the stator rotating magnetic field, so as to change the motor speed. By the formula $n = 60f/p$, it can be seen that the synchronous speed of asynchronous motor is inversely proportional to the number of pole pairs. The number of pole pairs is doubled, the synchronous speed n is reduced to half the original speed, and the rated speed n of the motor is also reduced by about half, so changing the number of pole pairs can achieve the purpose of changing the speed of the motor. Since this method of speed regulation is a step-by-step one, speed cannot be adjusted smoothly, and it is only suitable for squirrel cage motors.

Two-speed motors are mainly used in coal mines, petrochemical and chemical industries. In addition, it is widely used in textile, metallurgy, urban gas, transportation, grain and oil processing, paper making, medicine and other sectors. As the main power equipment, two-speed motors are usually used to drive pumps, fans, compressors and other transmission machinery.

With the development of science and technology, there are more and more places at the explosion risk. For example, edible oil production used to apply the traditional pressing process. After the 1970s, China began to introduce the advanced oil extraction process from abroad, but the chemical solvent containing ethane should be used in this process, ethane is a flammable and explosive substance. Therefore, the oil extraction workshop has become a dangerous place, requiring the use of two-speed motors and other explosion-proof electrical products. There is another example. In recent years, with the rapid development of China's highway, a large number of fuel gas stations have emerged, creating a new market for two-speed motors.

【Practice and Operation】

1. Task preparations

Prepare the required materials and components for the experiment according to the direct switching control circuit of the two-speed three-phase asynchronous motor.

2. Task contents

Connect the direct switching control circuit of two-speed three-phase asynchronous motor by Figure 2-30, and the experiment can only be carried out after the instructor checks the connection.

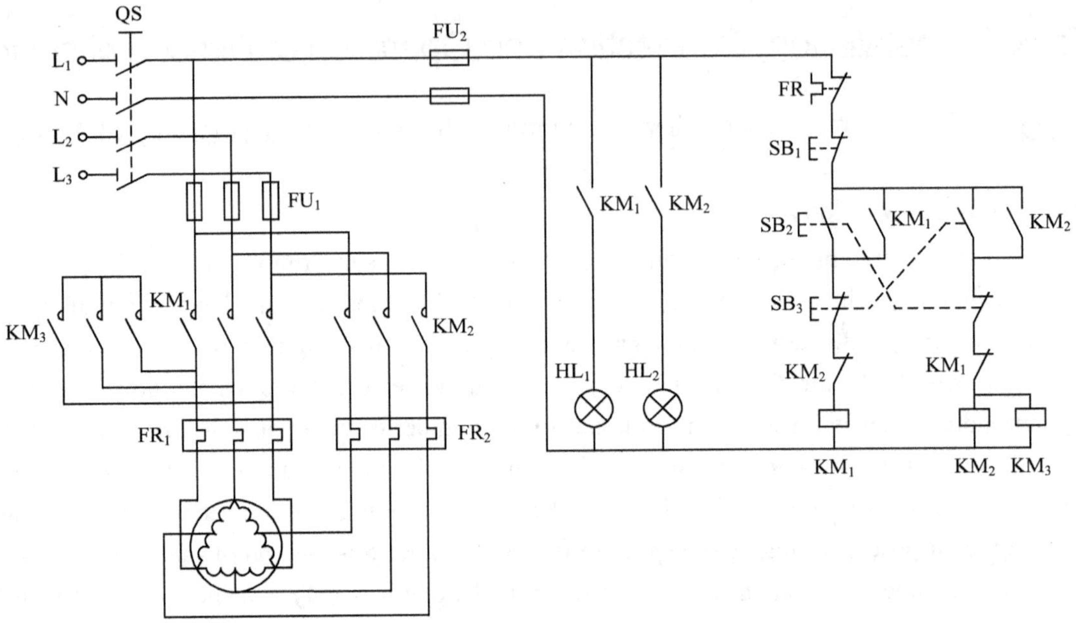

Fig. 2-30　Direct switching control circuit of two-speed three-phase asynchronous motor

3. Task procedures

(1) Switch on QS.

(2) Operate at a low speed by △ connection method:

Press SB_2 → KM_1 is powered on →
- KM_1 normally closed interlocking contact is disconnected
- KM_1 main contact is closed
- KM_1 self-locking contact is closed self-locking

→
- HL_1 is on
- The motor runs at a low speed by △ connection method

(3) Operate at a high speed by Y connection method:

Press SB_3 →
- KM_2 is powered on
- KM_3 is powered on

→
- KM_2 normally closed interlocking contact is disconnected
- KM_2 main contact is closed
- KM_2 self-locking contact is closed self-locking
- KM_3 main contact is closed

→
- HL_2 is on
- The motor runs at a high speed by Y connection method

(4) Pree SB_1 to stop the motor.

4. Precautions

Refer to the precautions in Task 3.

5. Notes and summary

____

____

____

____

____

【Extension Exercises】

Exercise 1 Connection of Braking Control Circuit of Half-Wave Rectification Energy Consumption of a Motor

1. Task objectives

(1) Have a grasp of the actual installation wiring by the schematic diagram for braking control circuit of half-wave rectification energy consumption of motor.

(2) Further understand the principle of half-wave rectification energy consumption braking through experiments.

2. Equipment

(1) Tools: screwdriver, electrician's pliers, wire stripper, needle-nosed pliers, measuring pencil, etc.;

(2) Meters: multimeter;

(3) Equipment required is shown in Table 2-4.

Table 2-4 Equipment required for skill training

Code	Name	Specification	Quantity
QS	Three-phase leakage switch	DZ47LE-32	1
FU_1	Fuse	RL1-15	3
FU_2	3P fuse	RT18-32	2
SB_1,SB_2	Button	LA19-11	2
FR	Thermal relay	JR36-20 0.3~0.5A	2
KM_1,KM_2	AC contactor	CJ20-10/220V	11
KT	Time relay	JS7-2A 220V	2

Continued

Code	Name	Specification	Quantity
M	Three-phase AC asynchronous motor		1
HL_1,HL_2	Indicator light	AD11-25/40	2
R	Variable resistance	BCI-150W	1
VD	Power diode		1
	Terminal board, trunking, conductor		Several

3. Task circuit

The installation and wiring should be carried out according to the half-wave energy consumption braking control circuit of the motor shown in Figure 2-31, and the electrification experiment should not be conducted until the wiring is checked and found correct by the instructor.

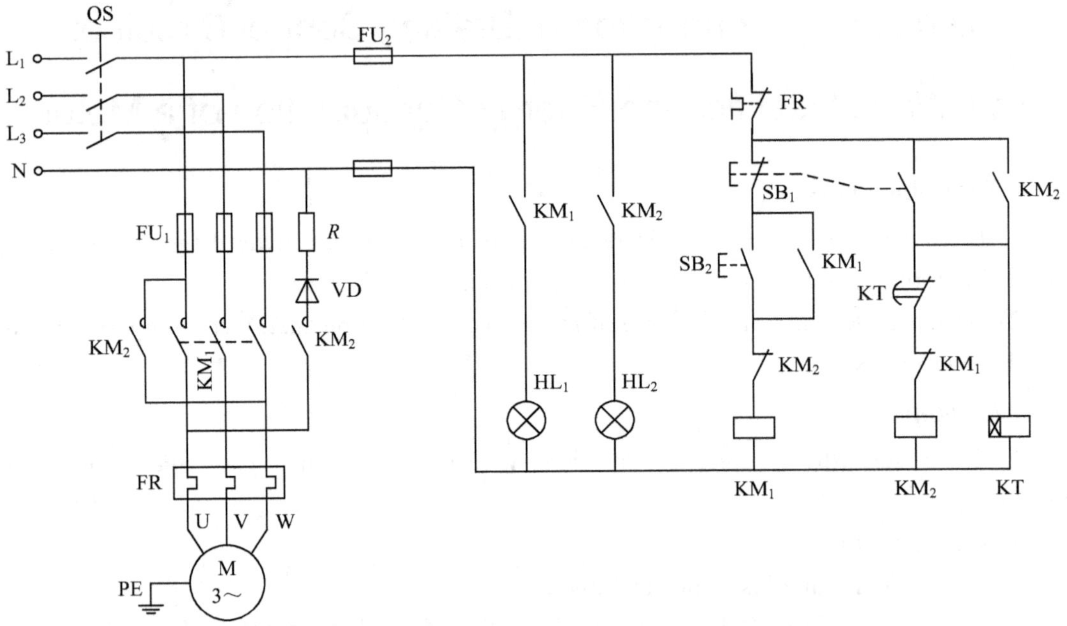

Fig. 2-31　Half-wave rectification energy consumption braking control circuit of the motor

4. Task procedures

(1) Turn QS on.

(2) Start the motor:

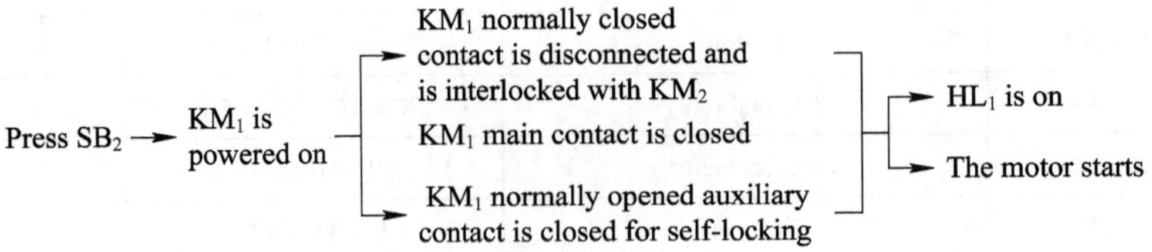

(3) Brake the motor:

Press SB_1 → KM_1 is powered off →
- → KM_1 main contact is disconnected to reset
- KM_1 normally opened auxiliary contact is disconnected and reset
- → KM_1 normally closed contact is closed to undo the interlocking with KM_2

→
- → HL_1 is off
- KT is on
- → KM_2 is on

→
- → KT delay starts at the normally closed contact
- KM_2 normally closed contact disconnects its interlocking with KM_1
- KM_2 main contact is closed
- → KM_2 normally opened contact is closed for self-locking

→
- → HL_2 is on
- → Motor stator windings are connected with DC current for energy consumption braking

→ KT delay at the normally closed contact is disconnected → KM_2 is off →
- → KM_2 main contact is disconnected and reset
- KM_2 normally opened auxiliary contact is disconnected and reset, HL_2 is off
- → KM_2 normally closed contact is closed to undo the interlocking with KM_1

→ The motor braking ends

Exercise 2 Connection of Full-Wave Rectification Energy Consumption Braking Control Circuit of a Motor

1. Task objectives

(1) Understand and master the principle of full-wave rectification energy consumption braking control circuit.

(2) Master the connection and operation method for full-wave rectification energy consumption braking control circuit.

2. Equipment

(1) Tools: screwdriver, electrical pliers, stripper pliers, needle-nosed pliers, test pencil, etc.;

(2) Meters: multimeter;

(3) Equipment required is as shown in Table 2-5.

Table 2-5　Equipment needed for skill training

Code	Name	Model	Quantity
QS	Three-phase leakage switch	DZ47LE-32	1
FU_1	Fuse	RL1-15	3
FU_2	3P fuse	RT18-32	1
FU_3	Fuse	RL1-15	1
SB_1, SB_2	Button	LA19-11	2
FR	Thermal relay	JR36-20 0.3～0.5 A	1
KM_1, KM_2	AC contactor	CJ20-10/220 V	2
KT	Time relay	JS7-2A 220 V	1
M	Three-phase AC asynchronous motor		1
HL_1, HL_2	Indicator light	AD11-25/40	2
R	Variable resistor	BCI-150W	1
TC	Transformer	380V/110V	1
VD	Rectifier bridge		1
	Terminal board, slot and wire		Suitable numbers

3. Task procedures

1) Be familiar with the principle of control circuit

For motors of a power below 10 kW, the control circuit of non-transformer and half-wave rectification energy consumption braking can be adopted. For motors of a power above 10 kW, energy consumption braking mostly adopts full-wave rectification energy consumption braking with a transformer.

The full-wave rectification energy consumption braking control circuit controlled by the time relay is shown in Figure 2-32. As can be seen from the main circuit, when the main contact of the Contactor KM_2 is closed, the 380 V AC power supply changes in voltage through the transformer and is rectified by the bridge rectification circuit to become a direct current, which flows through the stator winding and forms a constant magnetic field in the stator core. The variable resistor R is used to adjust the current through the stator winding and change the braking time.

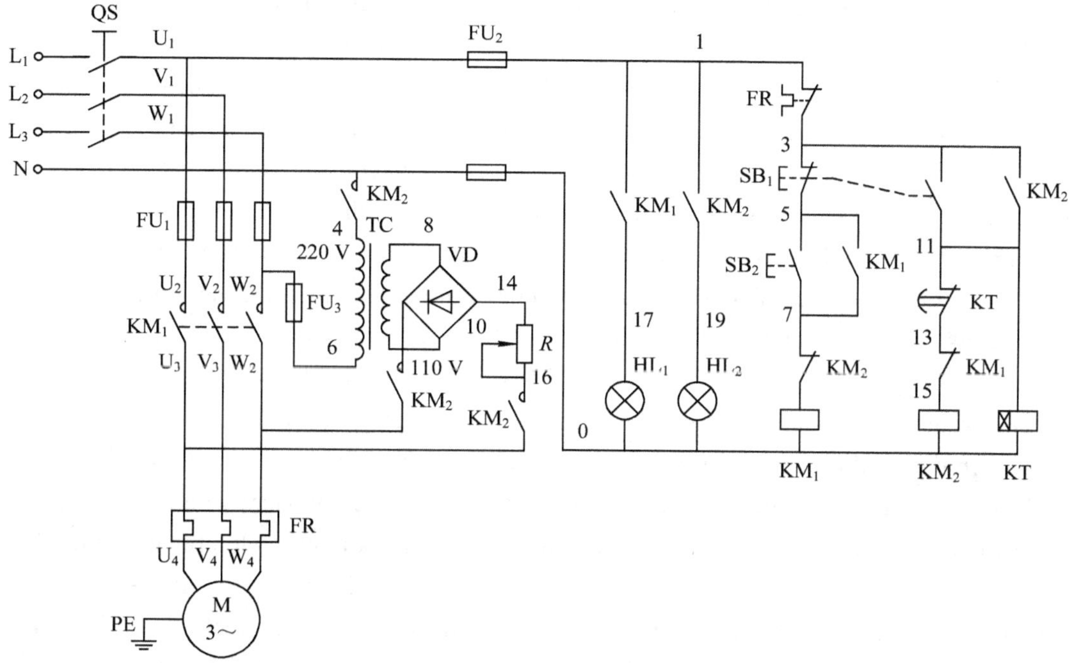

Fig. 2-32 Full-wave rectification energy consumption braking control circuit controlled by time relay

The working principle of the auxiliary circuit is as follows: switch on the power switch QS, press the button SB_2, the contactor KM_1 electromagnetic coil is energized, the main contact of KM_1 is closed, the motor M is started and operated, KM_1 normally opened auxiliary contact is closed to form self-locking, and the indicator HL_1 is on. For the motor to stop, press the button SB_1, the contactor KM_1 electromagnetic coil is cut off, the main contact of KM_1 is disconnected, and the motor M runs by inertia without power. At the same time, the coils of the contactor KM_2 and the time relay KT are electrically engaged, the main contact of KM_2 is closed, the normally opened contact of KM_2 is closed, the HL_2 indicator is on, and the DC current of the motor's M stator winding after full-wave rectification is used for energy consumption braking. The time relay KT starts timing. When the action time set by the time relay arrives, the normally closed contact by KT delay is disconnected, the contactor KM_2 electromagnetic coil is cut off power, and the main contact of KM_2 is disconnected from the DC power supply of full-wave rectification, and the energy consumption braking ends.

The DC voltage U_Z and DC current I_Z required by energy consumption braking can be calculated by the following two estimation formulas:

$$U_Z = I_Z R \qquad I_Z=(3.5\text{-}4)I_0 \text{ or } I_Z = 1.5\, I_N$$

R is the cold resistance at both ends of the stator winding applied by D C voltage, i.e. resistance at 15℃(Ω); I_0 is the line current of the motor with a load(A); I_N is the rated current of the motor (A).

2) Connect and check the lines according to the wiring diagram

According to the rated current of the motor, make a choice of the wires connecting the main circuit. Start from the QS in connection(note: this circuit uses three-phase four-wire system, in which N represents the middle line, while L_1, L_2, and L_3represent the fire line, also known as the phase line).We should first connect the main circuit, and then connect control circuit (this is the connection method for the simple circuit, but for the complex circuit we generally first connect the control circuit and then connect the main circuit to avoid the cross between wires).

First connect the three-phase leakage switch QS and the fuse FU_1 and KM_2 main contact to form circuits line U_1, V_1, and W_1.Then connect the fuse FU_1 and the contactor KM_1 main contact, and connect the fuse FU_3 and U_2, V_2, W_2 and KM_2 main contact. The fuse FU_3 is connected to the transformer primary side. Connect the transformer secondary side and the bridge rectifier VD (note: when connecting the transformer and the rectifier bridge, pay attention to the direction to avoid the connection in the opposite direction, or otherwise the component would be burned out. The rectifier bridge and the transformer are connected to the hanging plate 2(as in Figure 2-33, the rectifier bridge is connected to the numbered 8, 10, 12, and14 terminals, while the transformer is connected to the numbered 4, 6, 8, and 10 terminals). Rectifier bridge VD is connected with KM_2 main contact, and the adjustable resistance *R* respectively. The other side of the adjustable resistance *R* is connected to the one side of KM_2 main contact. The other side of KM_1 and KM_2 main contact is connected to the thermal relay.The other side of the thermal relay

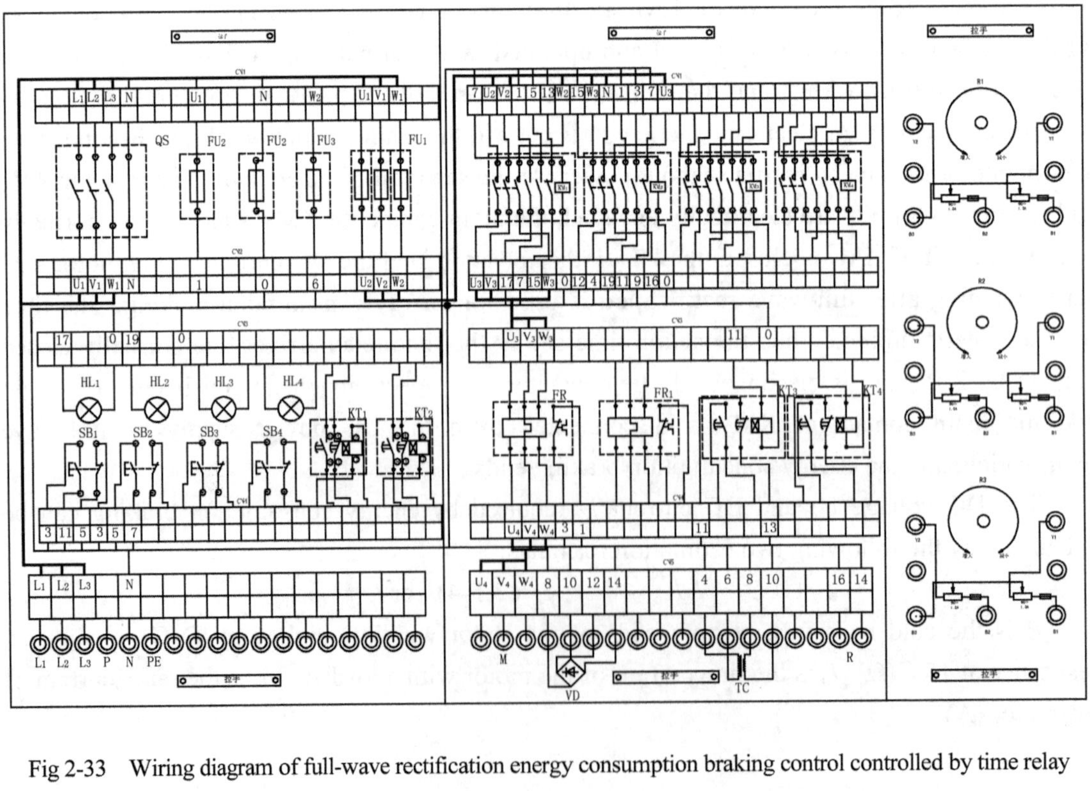

Fig 2-33　Wiring diagram of full-wave rectification energy consumption braking control controlled by time relay

is connected to U_4, V_4 and W_4 of the three-phrase asynchronous motor M. After the main circuit is connected, measure the required length between the fixed electrical components and intercept the appropriate length of the wire according to the wire number shown in the Figure 2-33. Remove the insulation layer at both ends of the wire, and connect the wire directly to the wiring terminal after the ends are covered in numbered wire tubes. After the connection of the auxiliary circuit, the metal shell of all the electrical appliances in the circuit must be installed with grounding wires.

As is discussed above, we have connected all the circuits. However, the coil of the AC contactor is 220 V, the control circuit is powered by 220 V, and the main circuit is powered by 380 V. Therefore, this circuit is powered by the three-phase four-wire system.

Upon completion of wiring installation, inspection must be carried out. Firstly, check whether there is insulation layer pressed into the wiring terminal, then check whether the total width of the exposed wire core exceeds the required value, and finally check the contact between all wires and the wiring terminals. Move and pull the wires on the terminals by hand. No loose wires are allowed. Then use the multimeter "R×10" to check the circuit. Take the two fuses of the control circuit as the measuring points, and first check whether there is a short circuit. Then press the start button, the resistance value indicated by the multimeter is the resistance value of the electromagnetic coil of the contactor, indicating that the control circuit is switched on when the start button is pressed. When the start button is released, the indicator of the multimeter returns to the position of "∞". After checking and making sure the circuit is correct, it is allowed to switch on the power.

3) Electrification test

After ensuring the connection is correct, test must be carried out under the supervision of the teacher.

Exercise 3 Motor Reverse Brake Control Circuit Connection

1. Task objectives

(1) Have a grasp of the actual installation wiring by the schematic diagram for the actual installation of the circuit wiring.

(2) Further understand the principle of the reverse connection braking of the motor through experiments.

2. Equipment

(1) Tools: screwdriver, electrician's pliers, wire stripper, needle-nosed pliers, measuring pencil, etc.;

(2) Meters: multimeter;

(3) Equipment required is as shown in Table 2-6.

Table 2-6 Equipment required for skill training

Code	Name	Specification	Quantity
QS	Three-phase leakage switch	DZ47LE-32	1
FU_1	Fuse	RL1-15	3
FU_2	3P fuse	RT18-32	2
SB_1, SB_2	Button	LA19-11	2
FR	Thermal relay	JR36-20 0.3~0.5A	1
KM_1, KM_2	AC contactor	CJ20-10/220V	2
M	Three-phase AC asynchronous motor		1
HL_1, HL_2	Indicator light	AD11-25/40	2
R	Variable resistance	BCI-150W	1
KS	Speed relay		2
VT	Thyristor		1
	Terminal board, trunking, conductor		Several

3. Task circuit

Install and connect the reverse connection braking control circuit of the motor as shown in Figure 2-34. The experiment can be powered on only after the instructor has checked it.

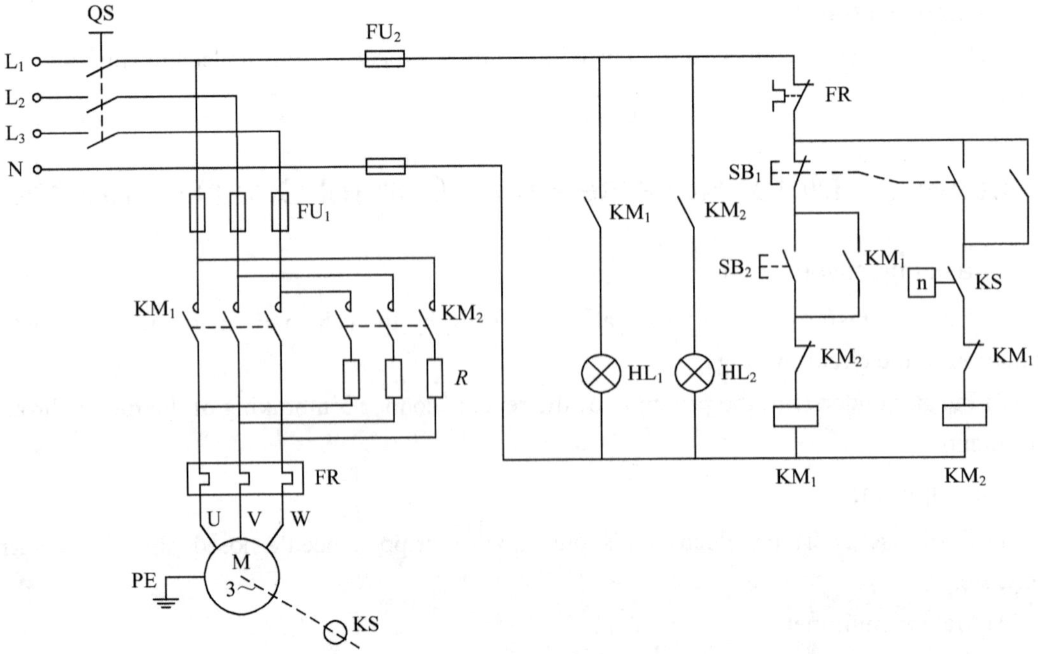

Fig. 2-34 Reverse connection braking control circuit of the motor

4. Task procedures

(1) Turn QS on.

(2) Start the motor:

Press SB_2 → KM_1 is powered on →
- KM_1 normally closed contact is disconnected to interlock with KM_2
- KM_1 main contact is closed
- KM_1 normally opened auxiliary contact is closed for self-locking

→ HL_1 is on; The motor starts

KS speed relay operates when the motor speed rises to 120 rpm

→ KS normally opened contact is closed (to prepare for stopping braking)

(3) Reverse brake the motor:

Press SB_1 → KM_1 is powered off →
- KM_1 main contact is disconnected and reset
- KM_1 normally opened auxiliary contact is disconnected and reset
- KM_1 normally closed contact is closed to undo the interlocking with KM_2

→ HL_1 is off; KM_2 is on →
- KM_2 normally closed contact is disconnected to the interlocking with KM_1
- KM_2 main contact is closed
- KM_2 normally opened auxiliary contact is closed for self-locking

→ HL_2 is on; Series resistance reverse brakes the motor

KS speed relay is reset when motor speed drops to 120 r/m

→ KS normally opened contact is disconnected → KM_2 is powered off →
- KM_2 main contact is disconnected and reset
- KM_2 normally opened auxiliary contact is disconnected and reset and HL_2 is off
- KM_2 normally closed contact is closed to undo the interlocking with KM_1

→ The motor braking ends

Exercise 4 Connection of Automatic Round-Trip Control Circuit of the Motor

1. Task objectives

(1) Understand and master the principle of automatic round-trip control circuit of the motor.

(2) Master the the connection method for automatic round-trip control circuit.

2. Equipment

(1) Tools: screwdriver, electrical pliers, wire stripping pliers, needle-nosed pliers, etc.;

(2) Meters: multimeter;

(3) Equipment required is as shown in Table 2-7.

Table 2-7　List of automatic round-trip control circuit components

Code	Name	Model	Quantity
QS	Three-phase leakage switch	DZ47LE-32	1
FU_1	Fuse	RL1-15	3
FU_2	3P fuse	RT18-32	1
KM_1, KM_2	Contactor	CJ20-10	2
SB_1, SB_2	Button	LA19-11	2
HL_1, HL_2	Indicator light	AD11-25/40	2
SQ_1, SQ_2, SQ_3, SQ_4	Travel switch	LX19-001	4
M	Three-phase AC asynchronous motor		1
FR	Thermal relay	JR36-20	1
	Terminal board, slot and wire		Suitable numbers

3. Task circuit

The automatic round-trip control circuit of the motor is as shown in Figure 2-35.

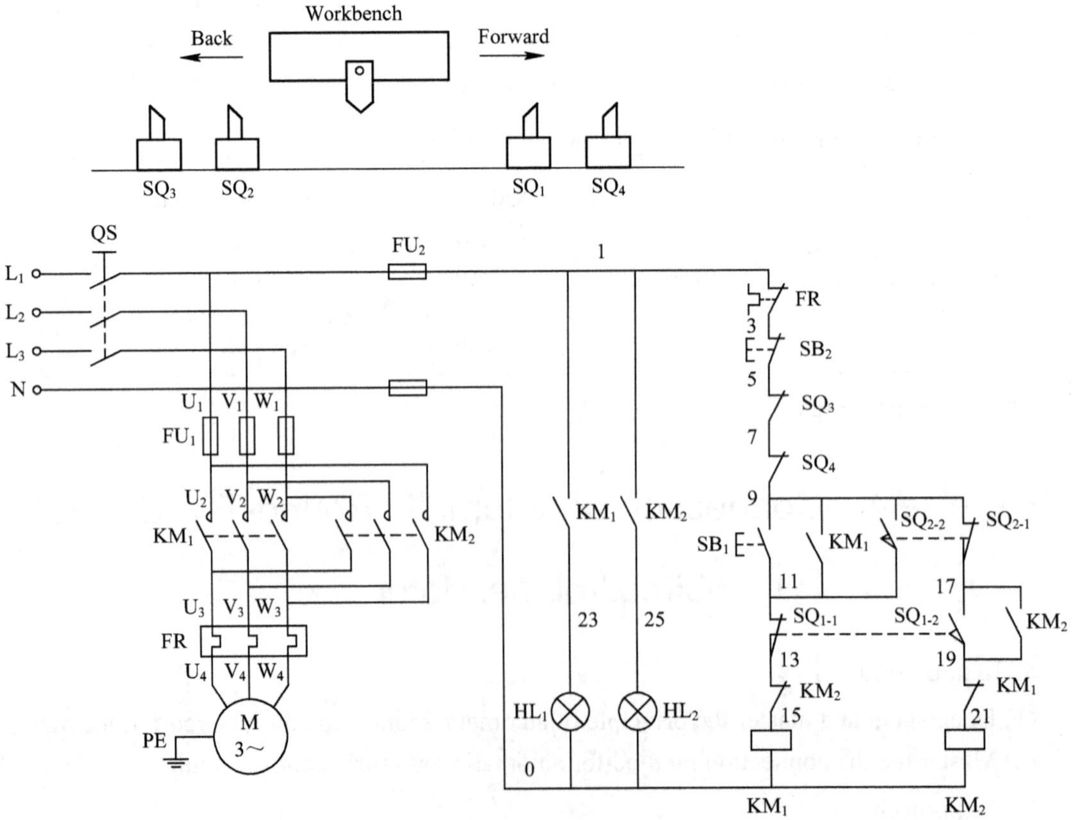

Fig. 2-35　Automatic round-trip control circuit of the motor

4. Task procedures

1) Be familiar with the principle of automatic round-trip control circuit

The automatic round-trip control circuit is shown in Figure 2-35. In the main circuit, when the contactor KM_1 is closed, the motor turns forward, while when the contactor KM_2 is closed, the motor reverses. The thermal elements of the thermal relay are connected in series in the motor circuit, and the working current of the motor is transmitted. Press the button SB_1, the electromagnetic coil of the contactor KM_1 is energized, the main contact is closed, and the positive rotation of the motor drives the moving parts of production machinery to move forward; KM_1 normally closed auxiliary contact is disconnected to interlock with KM_2; KM1 normally opened auxiliary contact is closed for self-locking; HL_1 indicating light is on.

When the moving parts of the production machinery reach the predetermined forward position, press down the stroke switch SQ_1 for SQ_{1-1} to disconnect, the contactor KM_1 electromagnetic coil is cut off; the main contact is disconnected; the auxiliary contact is reset; the indicator lamp HL_1 is off; the motor stops rotating. SQ_{1-2} is closed; the contactor KM_2 electromagnetic coil is energized; the main contact is closed; the motor inverts to drive the moving parts of production machinery to move backward. KM_2 normally closed auxiliary contact is disconnected to realize the interlocking with KM_1; normally opened auxiliary contact is closed for self-locking; HL_2 indicator is on. When the moving parts of the production machinery move backward, release the stroke switch SQ_1 for SQ_{1-1} to be reclosed and for SQ_{1-2} to be disconnected again.

When the moving parts of the production machinery move backward to the predetermined position, press down the stroke switch SQ_2 for SQ_{2-1} to disconnect; the contactor KM_2 electromagnetic coil is cut off; the main contact is disconnected; the motor stops rotating; the auxiliary contact is reset; the indicator lamp HL_2 is off. SQ_{2-2} is closed; the electromagnetic coil of the contactor KM_1 is energized; the main contact is closed; the positive rotation of the motor drives the moving parts of production machinery to move forward. The normally opened auxiliary contact is closed for self-locking; HL_1 indicator is on. The normally closed auxiliary contact is disconnected to realize the interlocking with KM_1. The cycle is repeated. SQ_3 and SQ_4 play a protective role to prevent the workbench accidents due to the unexpected failure in SQ_1 and SQ_2.

2) The installation wiring diagram of electrical components

The installation wiring diagram for the automatic round-trip control circuit is shown in Figure 2-36. Select appropriate wires and connect the automatic round-trip control circuit according to the installation wiring diagram. After each connecting wire is formed, the two ends of the wire shall be covered with a numbered tube.

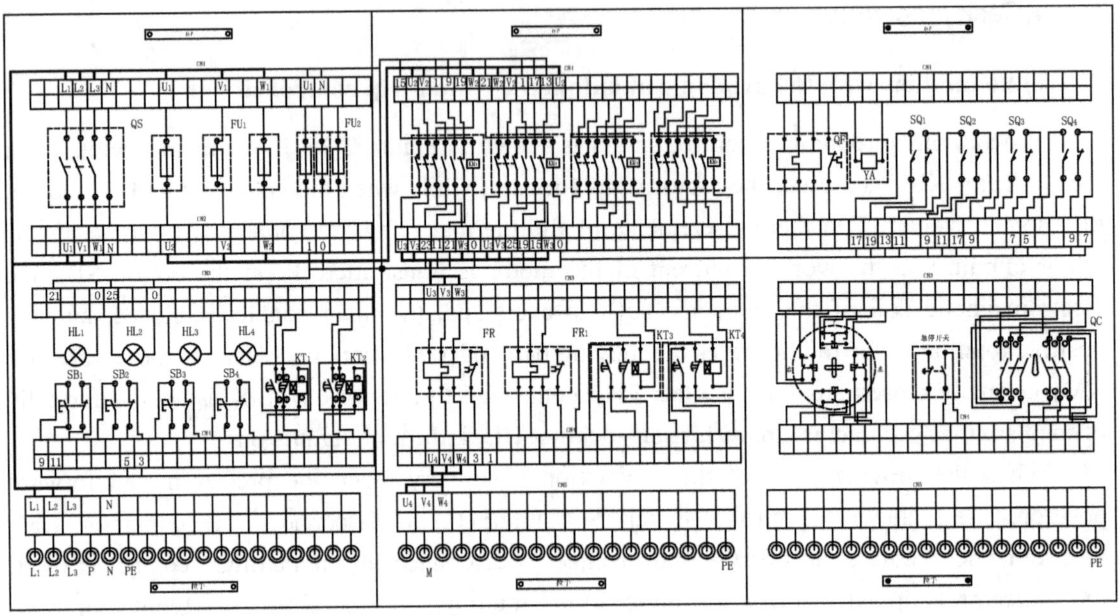

Fig. 2-36　Installation diagram for automatic round-trip control circuit

3) Check the wiring

After the completion of the connection, first check the connection to see whether there is missed and wrong connection; then check whether the wiring is firm and reliable; finally, check the circuit with a multimeter.

Select "R ×100" or "R×10" for the multimeter, and contact the two measuring pens with the two FU_2 fuses to see if there is a short circuit in the circuit. When there is no short circuit, press the button SB_1 and the pointer is deflected. The resistance value indicated by the multimeter is the resistance value of the solenoid coil of the contactor. Loosen SB_1 and press SQ_1, and the pointer is deflected. The reading is the same as the resistance of the electromagnetic coils of the contactors KM_1 and KM_2, indicating that the normally closed contacts in the electromagnetic coil branches of KM_1 and KM_2 are closed well, that the normally closed contacts are closed well and that the buttons are correct for the control circuit of the electromagnetic coils of the contactors.

Then check the self-locking and interlocking. The two measuring pens of the multimeter are still in contact with the two FU_2 fuses, press the auxiliary contact of the AC contactor, the multimeter indicates a deflection, and the resistance value indicated by the multimeter is the DC resistance of the electromagnetic coil of the AC contactor, indicating the self-locking. Press and hold the moving contact of the contactor KM_1 without releasing, and gently press the moving contact of KM_2, and the pointer of the multimeter returns to "∞", indicating that the normally closed contact of KM_2 is interlocked with KM_1. Use the same method to check the self-locking of KM_2 and the interlocking of SB_1 and KM_1 with KM_2. Press the moving contact of the contactor KM_1, the pointer is deflected and the reading of the multimeter is the same as the resistance of the electromagnetic coil of the contactor KM_1. When the moving contact of the contactor KM_1 is not released, the stroke switch SQ_1 is pressed down. The pointer of the multimeter first returns to "∞" and then is deflected. The reading of the multimeter is the resistance of the electromagnetic coil of the contactor KM_2, indicating that the stroke switch SQ_1 can switch the rotation direction of the motor. In the same way, check whether the function of the stroke switch SQ_2 is correct, but it should be noted that the moving contact of the contactor KM_2 should be pressed at this time and the stroke switch SQ_2 is pressed down. SQ_3 and SQ_4 in this circuit are limit protection at both ends, replaced by SB_3 and SB_4 normally closed contacts.

4) Inspection with power on

After the check by multimeter, the power can be turned on under the supervision of the instructor. When the power is on, the correct actions of the electrical appliances in the circuit are: close the power switch QS, press the button SB_1, the contactor KM_1 is closed and the motor moves forward; when SQ_1 is pressed down, the contactor KM_1 is disconnected and KM_2 is closed; when SQ_2 is pressed down, the contactor KM_2 is disconnected and KM_1 is closed down; when SB_2 is pressed down, SQ_3 and SQ_4 contacts can all be disconnected. It shows that the moving parts of the production machinery can move back and forth. When the circuit is checked to be correct, making sure that there are no missing phase errors, the motor will be connected for the test with power on.

5) Test with power on

After the connection is checked to correct, test with power on must be carried out under the supervision of the teacher.

Exercise 5 Connection of Sequential Start Control Circuit Connection of the Motor

1. Task objectives

(1) Master the skills of the actual installation of the circuit by the schematic diagram for sequential start control circuit of the motor.

(2) Deepen the understanding of the working principle of the circuit through experiments.

2. Task circuit

Connect the wires according to Figure 2-37. The experiment is carried out after the instructor's check of the wiring.

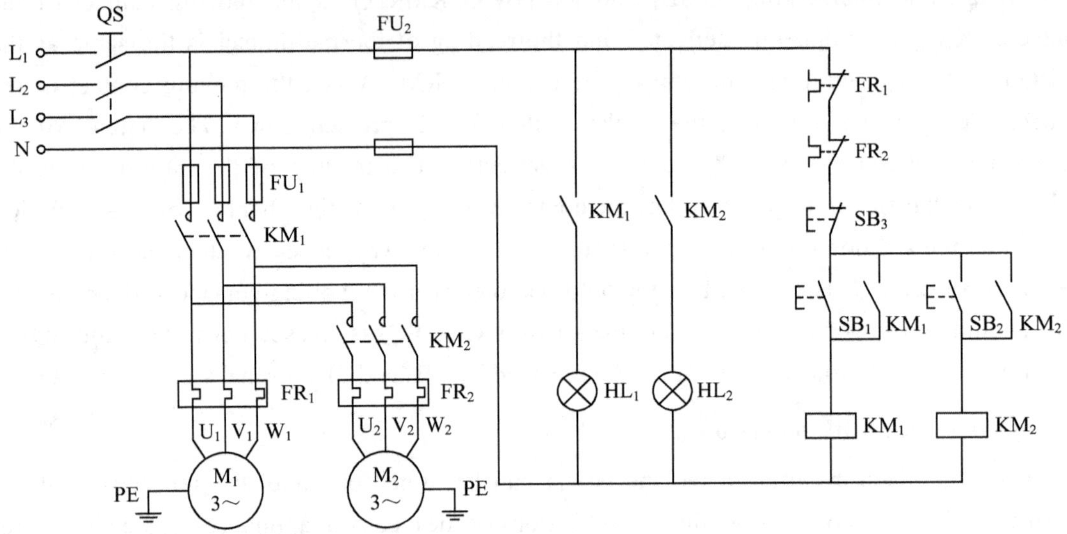

Fig. 2-37　Motor sequential start control circuit of the motor

3. Task procedures

(1) The motor M_1 starts:

Close the switch QS → Press SB_1 → KM_1 is on →
- KM_1 main contact is closed → The motor M_1 operates
- KM_1 normally opened self-locking contact is closed → HL_1 is on

(2) the motor M_2 starts:

Press SB_2 under the action of M_1 → KM_2 is on →
- KM_2 main contact is closed → M_2 is on and operates
- KM_2 self-locking contact is closed → HL_2 is on

(3) The motors stop:

Press SB_3 → KM_1 and KM_2 are off →
- KM_1 and KM_2 main contacts are disconnected → The motors M_1 and M_2 stop
- KM_1 and KM_2 normally opened contacts are disconnected → HL_1, HL_2 are off

Note: the motor M_2 can't start without the motor M_1 running.

Module Three Comprehensive Training in Analog Electronic Technology

【General Introduction】

This project has designed three typical tasks of basic electronic circuit installation and adjustment test for elementary-level electricians. The main training objectives of these tasks are the cognition of principles, simple measurement, circuit welding training and more complex circuit making. The training involves the performance test of semiconductor diode and triode, the basic training of electronic circuit welding, and the making of DC voltage stabilizing power supply. Trainees can meet the basic requirements for basic skills of electronic circuits for elementary electricians through the practical training of this project.

【Basic Elements】

1. Knowledge objectives

(1) Understand the characteristics and working principle of semiconductor devices.

(2) Understand the composition and working principle of DC voltage stabilizing circuit.

(3) Master the electronic welding process.

(4) Master the classification and selection of electric soldering iron and welding wire as well as the selection of flux.

2. Competence objectives

(1) Capable of making and measuring half-wave and full-wave rectification and voltage stabilizing circuits.

(2) Capable of choosing welding tools for welding objects.

(3) Capable of carrying out pre-welding treatment.

(4) Capable of installing and welding resistance, capacitance, diode and other basic components of the simple circuit.

(5) Capable of identifying false welding.

3. Operation standards

1) Preparation before the task

(1) Read the experimental instructions carefully, and make notes according to the relevant contents of the relevant teaching materials. Do your homework in advance.

(2) Be aware of the experimental purpose, list the experimental items and procedures, and go to the laboratory to understand the structural characteristics of the experimental devices,

experimental units and required components and instruments.

(3) Keep in mind the matters needing attention in the experiment to avoid personal safety accidents and equipment damage.

2) Task procedures

(1) Establish a group and divide the work reasonably. The practical training task is carried out in groups; each group is composed of 2-3 people. The task usually includes the selection of tools, wiring production, instrument measurement, test data recording and other work. The members of the experimental team should have a clear and reasonable division of labor, take positions and coordinate their work. In addition, they should carry out continuous consultations in the project to ensure the standard and coordination of the practical training.

(2) Select units and instruments. Select appropriate operation units and instruments (including range grade of instruments) according to the purpose of the training project, and record their models and parameters.

(3) Wiring and welding according to the drawing. Carry out the electronic circuit connection or the welding production of the project circuit according to the circuit diagram in the practical training.

(4) Reading and recording. Pay attention to multiplier, minimum scale value, unit and significant figure when reading an instrument. If anything abnormal is found, cut off the power supply immediately, find out the cause and troubleshoot the fault. After the data test, check whether the data are correct according to the working principle of the project circuit. Line connecting or welding problems should be checked if they differ from theoretical expectations.

(5) Removing wires. The power supply must be turned off after the completion of the task, and then wait for a little time, so that capacitors and other energy storage elements discharge. Then the wires can be removed, and the wire, instrument and circuit unit module are well arranged, and the soldering iron must be cooled to normal temperature after the power is cut off and be put away.

3) Training report

At the end of each program, students are required to write a training report. The report should highlight the process and contents of work tasks and record the measurement data or the function of task circuit. Problems found in the process of practical operation should be timely recorded in the report for analysis and solution. The practical training report is not only the record of practical training operation process, but also an important link of learning, discovering and solving problems. It can comprehensively improve students' logical ability, graphic and text processing ability and data analysis ability. The report requires clear thinking, concise writing, neat charts and graphs, reasonable analysis, clear conclusions, and standard writing.

4. Safety standards

(1) Ensure that the training device has reliable grounding protection, leakage protection, overload protection, insulating table top and carpet and safety connector wire.

(2) Human body should not contact the exposed charged body during the implementation of the task.

(3) The power supply must be cut off when wiring and unwiring.

(4) The system can be powered on after the line check by the instructor and other team members.

(5) In case of abnormal conditions, such as smelling peculiar smell (such as the burnt smell of varnish), finding components or equipment overheating or excessively vibrating (or excessive running noise), fuse burnt out, sparks in the unit, or seeing the meter over the full range (or reverse deviation), we should immediately cut off the power supply, and then find the cause and troubleshoot the problem.

(6) Be familiar with the board drawing of the welded PCB, and check whether the model, specification and quantity of components meet the requirements in the drawing, and report the problems in time.

(7) Check whether the pins or leads of various components are over-oxidized and whether the component identification are unclear. In case of the above problems, it must be selected for other treatment. Check the board for deformation and deflection.

(8) The welding sequence of components is based on the principle: first low, and then high; first small, and then large. In general, welding should be carried out in the sequence of resistance, capacitance, diode, triode, integrated circuit and high-power tube.

5. Quality standards

(1) Components should not be assembled wrongly, unloaded, connected wrong or askew loose, etc..

(2) Solder joints should be full of tin with no burr, pinhole, bubble, crack, hanging tin, drawing point, lack of welding, butt welding, false welding and other defects.

(3) The metal surface of the connected circuit board should be free from corrosion and other impurities.

(4) The soldered circuit board should not have spots, cracks, bubbles, whitening and other phenomena, and the copper foil and shaped coating layer should not fall off, warp or stratify.

(5) The surface of the pins or leads of components should be evenly permeated with tin.

【Training Requirements】

1. Preparation of tools, instruments and equipment

Before training, prepare required tools and equipment as shown in Table 3-1.

Table3-1 Tools and equipment required for comprehensive training in analog electronics technology

No.	Name	Model	Unit	Quantity
1	Diode		Piece	Several
2	Bipolar transistor	9013	Piece	Several
3	Field effect transistor	IRF630	Piece	Several

Continued

No.	Name	Model	Unit	Quantity
4	YL135 unit modules RP1, RP3, VT3, VT5, R01, and R15, DC voltage regulating power supply for training platform		Group	1
5	Multimeter		Piece	1
6	Electric iron		Piece	1
7	Tin solder		Plate	1
8	Resistor, capacitor and other electronic components			Several
9	Circuit board			Several
10	Wire and stripping pliers			Several

2. Knowledge assessment points

(1) Master the working principle and characteristics of diodes and transistors.

(2) Master the working principle of basic amplifier circuit and rectifier circuit.

(3) Master the electronic welding process and the method for the classification and selection of electric soldering iron and welding wire, as well as the method for the selection of flux.

3. Skill assessment points

(1) Be able to make correct use of a digital multimeter to test whether the diode is good or bad.

(2) Be able to install, and weld resistor, capacitor, diode and other basic components of a simple circuit.

(3) Be able to complete the fabrication and performance test of a voltage stabilizing circuit.

【Work Tasks】

Task 1　Performance Test of Semiconductor Diodes and Triodes

【Relevant Knowledge】

1. Intrinsic Semiconductor

Intrinsic semiconductors are pure semiconductors of crystal structure.

1) Crystal structure of semiconductors

Semiconductor refers to the material with conductivity between conductor and insulator at room temperature. Semiconductors are used in integrated circuits, consumer electronics, communication systems, photovoltaic power generation, lighting applications, high-power

power source conversion and other fields. For example, diodes are devices made of semiconductors. No matter from the perspective of science and technology or economic development, the importance of semiconductor is very huge. Most of today's electronic products, such as computers, mobile phones or digital recorders, have a very close relationship with semiconductors. Common semiconductor materials include silicon, germanium, and gallium arsenide. Silicon is the most influential semiconductor material in commercial applications.

The structure of a silicon atom is shown in Figure 3-1. There are four electrons in the outermost layer of a silicon atom. The common semiconductor material germanium also has the same crystal structure, that is, the outermost layer of the atom has four extranuclear electrons.

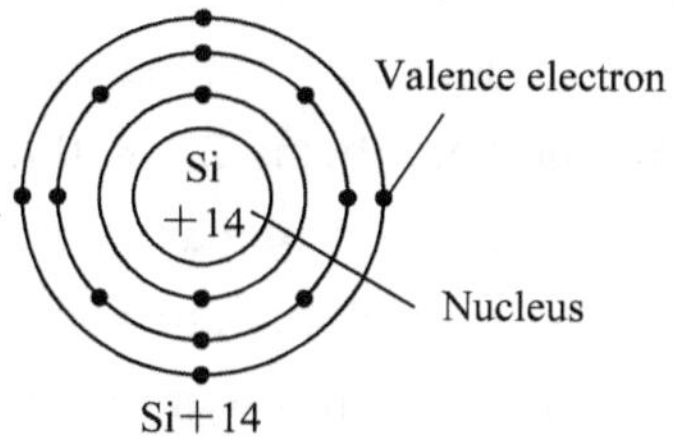

Fig. 3-1 Atomic structure of silicon

2) Two kind of carriers in intrinsic semiconductors

Due to the excitation of heat and light, electrons in intrinsic semiconductors will break away from covalent bonds and become "free electrons".

At the same time, a vacancy called "hole" is left in the original position.

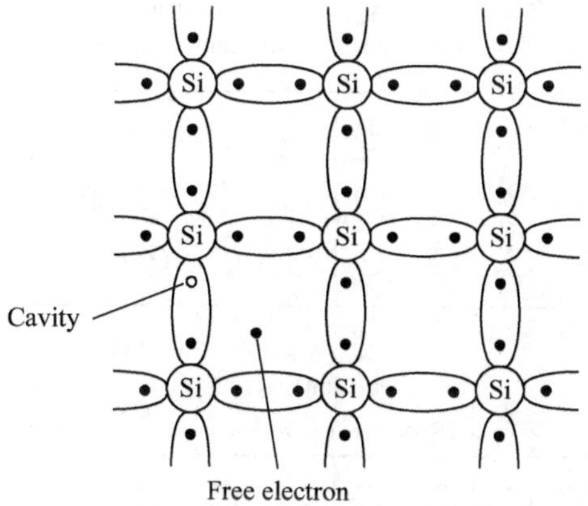

Fig. 3-2 Intrinsic semiconductor structure

3) Characteristics of intrinsic excitation

(1) The numbers of free electrons and the numbers of holes are equal, forming a pair of electron holes, and the concentration of free electrons and holes is about the same, of about 1.5 ×

$10^{10}/cm^3$.

(2) In the absence of electric field, the movement of electrons and holes is not directional and no current is formed.

(3) Under the action of electric field, holes move along the electric field, forming a hole flow; free electrons move against the electric field, forming an electron flow. Semiconductors have two kinds of carriers involved in conduction.

(4) The numbers of free electrons and the numbers of holes depend on temperature.

4) The PN junction

In intrinsic semiconductor, since there are few hole electron pairs, the conductivity is low. The addition of 3- or 5-valence elements helps to improve the conductivity of semiconductor.

(1) N-type semiconductor. N-type semiconductors are formed by adding traces of 5-valence elements into silicon or germanium. Multi-electrons are negatively charged, so they are called N-type semiconductors. Carrier concentration in N-type semiconductors is as follows: free electrons being 1.5×10^5-$1.5\times10^6/cm^3$ and holes being $2.3 \times 10^5/cm^3$.

(2) P-type semiconductor. P-type semiconductors are formed by adding traces of 3-valence elements into silicon or germanium. Multi-electrons are positively charged, so they are called P-type semiconductors. Carrier concentration in P-type semiconductors is as follows: holes being 1.5×10^5-$1.5\times10^6/cm^3$ and free electrons being $2.3 \times10^5/cm^3$.

(3) Formation of PN junction. As shown in Figure 3-3, the movement caused by concentration difference is called diffusion motion. Because of the internal electric field, the electrons in P region and the holes in N region move towards each other. Diffusion and drift balance form a space charge region (depletion layer). The depletion layer acts as a barrier to multi-carrier diffusion but promotes minority carrier drift.

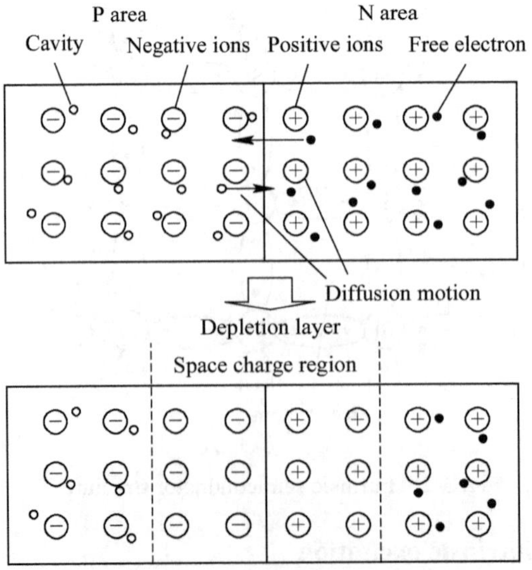

Fig. 3-3 Diffusion drift motion diagram of charges

(4) Unidirectional conductivity of PN junction:

① PN junction conducts by applying a forward voltage. The direction of the external electric field is opposite to that of the internal electric field. The outer electric field narrows the space charge region and facilitates diffusion. With many holes in P region and many electrons in N region, the applied voltage enables electrons to spread from N region to P region, the forward current is from P region to N region, and the PN junction is of low resistance and is turned on.

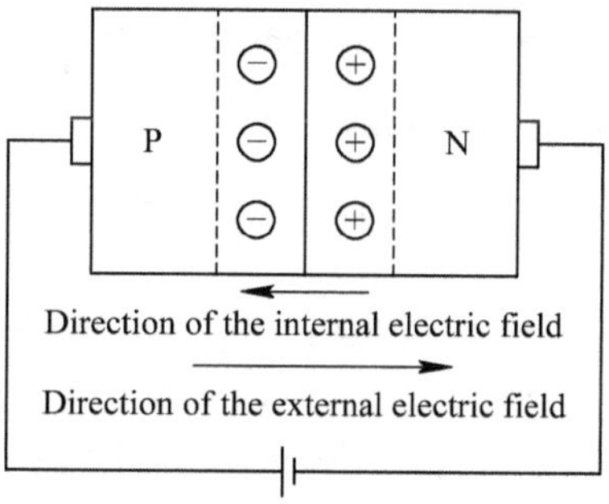

Fig. 3-4 PN junction forward conduction

② PN junction is cut off by applying a reverse voltage. The direction of the external electric field is the same as that of the internal electric field. The external electric field widens the space charge region, which makes it difficult to diffuse but easy to drift. The minority electrons in N region conduct to form a reverse saturation current, and PN junction is of high resistance, which is called reverse bias cut-off.

③ There are effects of temperature and capacitance on PN junction. The increase of temperature will lead to the rapid increase of reverse current, and the capacitance effect of diode must be considered in high- frequency circuit.

2. Working principle of diode

1) Working principle

The working principle of the diode is simply summarized as forward conduction and reverse cut-off the symbol for which is shown in Figure 3-5. Crystal diode is a PN junction formed by P-type semiconductor and N-type semiconductor. Space charge layer is formed on both sides of its interface and self-built electric field is formed. When there is no applied voltage, the diffusion current caused by the carrier concentration difference on both sides of the PN junction is equal to the drift current caused by the built-in electric field, so it is in the state of electric equilibrium. When there is a positive voltage bias, the interaction between the external electric field and the self-built electric field causes the increase of carrier diffusion current and the positive current. When there is a reverse voltage bias, the external electric field and the self-built electric field are further strengthened, and the reverse saturated current I_0 independent of the reverse bias voltage is formed within a certain range of reverse voltage. When the applied reverse voltage is high to a certain extent, the electric field intensity in the space charge layer of the PN junction reaches the critical value, resulting in a large number of electron hole pairs and a great reverse breakdown current, which is called diode breakdown phenomenon. The reverse breakdown of PN junction can be divided into zener breakdown and avalanche breakdown.

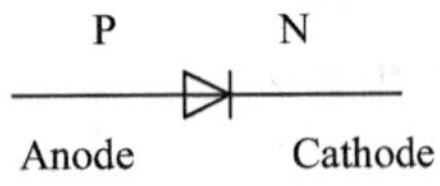

Fig. 3-5 Symbol for diode

2) Volt-ampere characteristics of diodes

(1) Forward conduction voltage drop: for a germanium tube the drop is about 0.3 V; for a silicon tube the drop is about 0.7 V.

(2) Forward conduction dead-time voltage: for a germanium tube, it is about 0.2 V; for a silicon tube, it is about 0.5 V.

(3) Reverse saturated current: it refers to current before reverse breakdown; saturated current increases rapidly with temperature.

(4) Reverse breakdown voltage: it refers to the voltage when the diode breaks down.

3) Main parameters for diodes

(1) Maximum rectifying current. It refers to the maximum forward average current allowed to flow through a diode. Its value is related to PN junction area and external heat dissipation conditions. If the average current through the diode exceeds this value for a long time, it will burn out because of overheating of PN junction.

(2) Maximum reverse operating voltage. It refers to the maximum operating voltage allowed by the diode.

(3) Reverse current. It refers to the current when the diode is not broken down.The smaller the reverse current, the better the unidirectional conductivity of the diode. Reverse current is very sensitive to temperature.

(4) Maximum operating frequency. It refers to the maximum operating frequency allowed by the diode. When the operating frequency exceeds this value, the unidirectional conductivity of the diode will not be well reflected. Its value depends on the capacitance of PN junction. The larger the capacitance is, the lower the frequency is.

3. Triode

Triode, also known as semiconductor triode, bipolar transistor, and crystal triode, is a kind of semiconductor device to control current. Its function is to amplify weak signals into electrical signals of larger amplitude, and it is also used as contactless switches. Triode is one of the basic components of semiconductor. It has the function of current amplification and is the core component of electronic circuit. The triode is made of two PN junctions which are very close to each other on a semiconductor substrate. The two PN junctions divide the whole semiconductor into three parts, the middle part being the base area, and the two sides being the emitter area and the collector area. There are two types of arrangement, PNP and NPN, as shown in Figure 3-6.

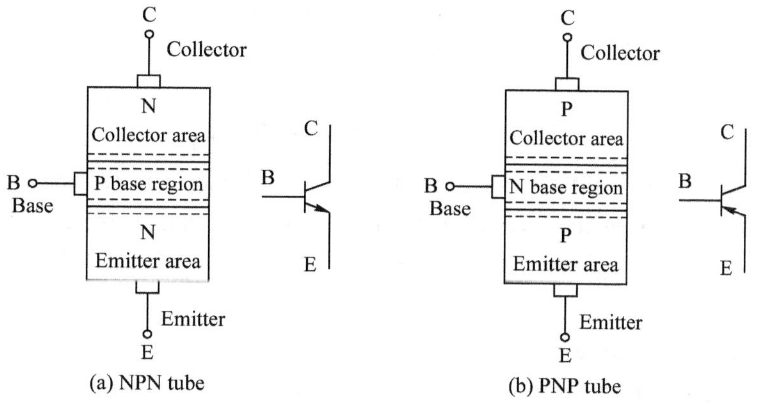

Fig. 3-6 Two kinds of transistor structure

Characteristics of base area, emitter area and collector area:

(1) The doping concentration in the emitter region is much higher than that in the base region, which is conducive to multi-carrier emission to the base region.

(2) The base region is very thin and less doped, which is conducive to the carriers passing through the base region and less recombined.

(3) The geometric size of the collector region is larger than that of the emitter region and the concentration is lower, which is beneficial to collecting carriers.

Therefore, the emitter and collector regions cannot be interchanged.

4. YL135 experimental platform

Yalong YL135 electronic technology training and assessment device is selected as the experimental platform. The +5 V DC regulated power supply on the test bench is used to supply power to the module of electronic component unit and 220 V AC power supply to the soldering iron.

【Practice and Operation】

1. Task preparations

(1) Preview task contents.

(2) Prepare the diode, and triode unit circuit module, multimeter, soldering iron, wire, soldering tin and other equipment needed for the task.

2. Task contents

1) Check the quality of diodes

The positive and negative switching of the diode is detected by the red and black multimeter pen, shown in Figure 3-7.

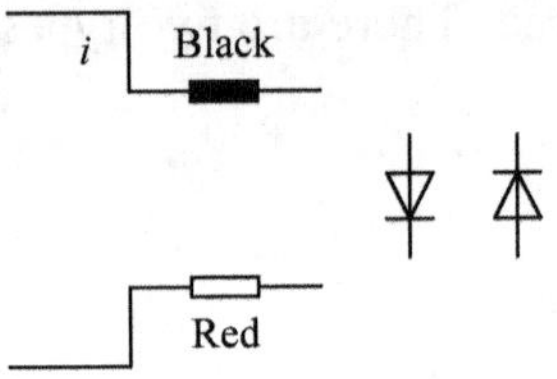

Fig.3-7　Test of diode by multimeter

Measure the forward and reverse resistance of the diode. The rectifier diode 1N4007 is selected by VD1 unit. Measure the forward and reverse resistance of the diode as shown in Figure 3-7. Record the $R_{d\ positive}$ = ______ Ω and the $R_{d\ negative}$ = ________k Ω, and judge whether the diode is normal or not.

2) Diode performance test

The performance test circuit of diode are shown in Figure 3-8. The volt-ampere characteristic curve is shown in Figure 3-9. Here, its forward volt-ampere characteristics are mainly measured, and the test data are filled in Table 3-2.

Adjust the current voltage from 0 to 2 V, and observe the reverse saturation current according to Figure 3-8.

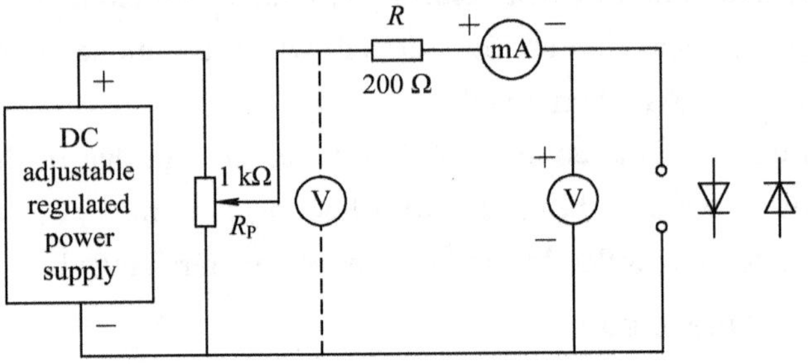

Fig. 3-8　Performance test circuit of diode

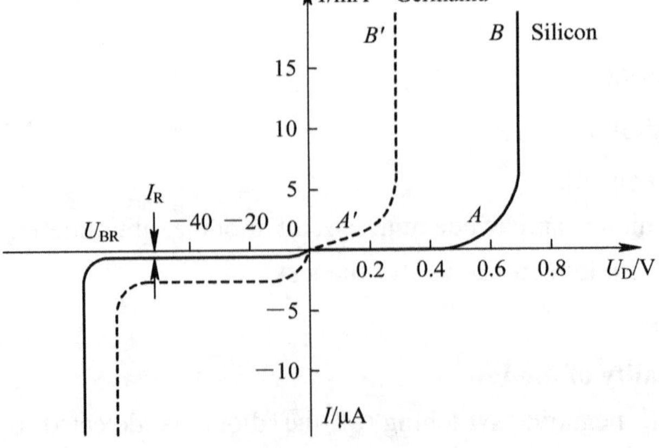

Fig. 3-9　Volt-ampere characteristic curve of diode

Table 3-2 Record of forward characteristic of diode

U_d/V	0	0.05	0.10	0.15	0.20	0.30	0.40	0.50	0.60	0.70
I_d/A										

For reverse volt-ampere characteristics, the reverse turning voltage (U_{BR}) is usually very high (for example, it is 1000 V for 1N4007). Therefore, only reverse leakage current I_R (also known as reverse saturation current) is measured here.

3) Test the output characteristics of triode

The main parameters for the NPN bipolar transistor 9013 are as follows:

(1) Maximum collector power P_{CM}: 400 mW.

(2) Maximum collector current I_{CM}: 500 mA.

(3) Collector-emitter breakdown voltage $U_{(BR)\,CEO}$: 25 V.

(4) Collector-emitter penetration current I_{CEO}: 0.5 mA.

(5) Collector-emitter saturation voltage $U_{CE\,(sat)}$: 0.6 V.

(6) Cutoff frequency F_T: 150 MHz.

(7) Current amplification β: 64-144.

The main parameters for the N-channel enhanced field effect tube are as follows:

(1) Source maximum current $I_{UD(max)\,GS}$: 9 A.

(2) Drain to source breakdown voltage $U_{DS(BR)}$: 200 V.

(3) Drain to source on resistance $R_{DS\,(on)}$: 0.4 Ω.

(4) Maximum dissipated power P_D: 80 W.

3. Task procedures

(1) Prepare diode, triode, multimeter, diode forward test unit and DC power supply.

(2) Use the red and black pen of the multimeter to measure the on-off of the diode, and set the resistance to "R ×100" or "R× 1k". The forward resistance of general diodes is hundreds of ohms, and the reverse resistance is hundreds of kilohms. If the forward and reverse resistances of diode are very small, it indicates that there has been a short circuit in the diode. If both the forward and reverse resistances are large, it indicates that the diode has been broken.

(3) Test the performance of the diode, wiring according to Figure 3-8. The forward of the diode is connected into the circuit, use YL-135 training platform DC adjustable power supply to apply 2 V voltage to the circuit, adjust the potentiometer R_P, measure the output voltage U_d, and fill in Table 3-2.

(4) In the above practical training, reverse the diode and replace the milliampere with the microampere, and adjust the power supply voltage U_d to 10 V. Measure the reverse saturation current I_R=_____μA.

(5) According to Figure 3-10, a test circuit is constructed, where E_b is 3 V DC power supply and E_c is DC adjustable regulated power supply. Fill in Table 3-3.

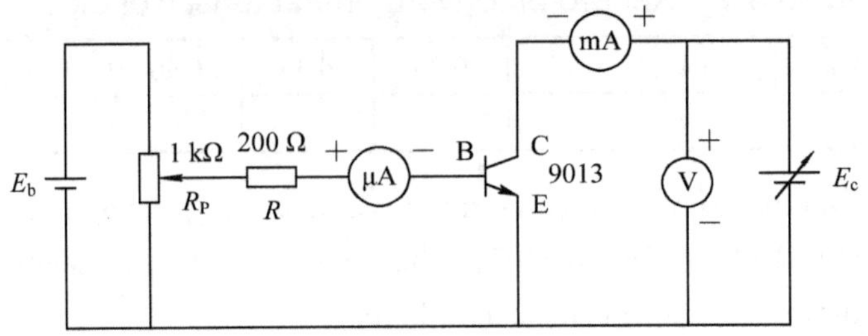

Fig. 3-10 Test Circuit for output characteristics of bipolar transistor

Table 3-3 Output characteristics of bipolar transistor(I_C/mA)

I_B/μA	U_{CE}/V					
	0	0.20	0.50	1.0	5.0	10
0						
20						
40						
80						
120						

(6) Complete wiring as shown in Figure 3-11, where E_G is supplied by a DC power source with different grades, and E_G are 4 V, 5 V, 6 V and 7 V respectively. E_D is powered by a DC adjustable and regulated power supply. E_D are 0 V, 3 V, 6 V, 9 V, 12 V, 15 V and 18 V respectively. Complete Table 3-4.

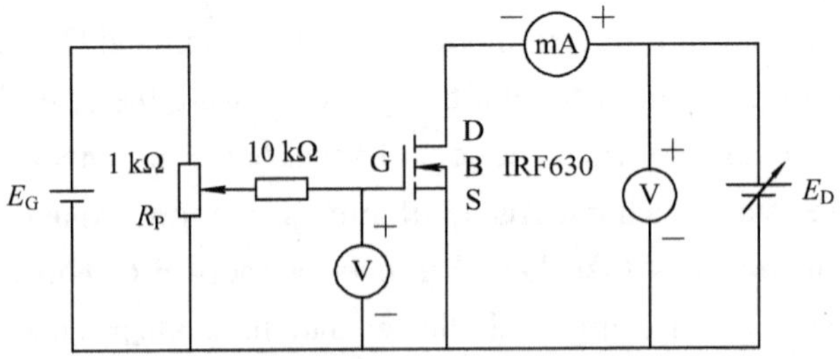

Fig. 3-11 Test Circuit for output characteristics of field effect tube

Table 3-4 Output characteristics of field effect tube(I_D/A)

U_{GS}/V	U_{DS}/V						
	0	3	6	9	12	15	18
4							
5							
6							
7							

4. Precautions

(1) Line connection checks must be carried out before the power supply operation test.

(2) In the process of power-on operation, if the multimeter is used to measure parameters, the gear must be switched to the correct and appropriate position. Since diode forward resistance is small, attention should be paid to applying current limiting resistance, in order to prevent excessively large current from burning out the diode.

(3) The power supply voltage regulating potentiometer should be set to the lowest voltage at the beginning to avoid over-high voltage.

(4) After the completion of the training, the cleaning up of the training places should be done well, and the training equipment should be withdrawn or put away in time.

5. Notes and summary

____

____

____

____

Task 2 Basic Training of PCB Welding

【Relevant Knowledge】

1. Basic introduction to and use of electric soldering iron

The soldering iron is one of the main welding tools. Before using the new soldering iron, apply a file to modify both sides of a big soldering head as shown in Figure 3-12. And sharpen the chamfer on the head of the soldering iron to prevent burrs from damaging the PCB soldering pad. If a long-life soldering iron head is used, no processing is required.

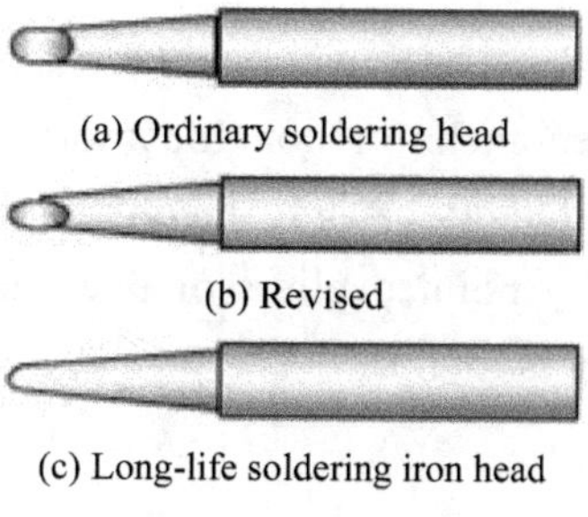

Fig. 3-12 Commonly used soldering heads

Soldering iron temperature and welding time should be appropriate. When welding, the soldering head should be heated to a temperature higher than the melting point of the solder, and weld for a proper time, generally within 3 seconds. Long-lasting time will make the printed circuit board copper upturned, damaging the circuit board and electronic components.

2. Basic welding methods

Generally, a solder of the diameter of 1.2-1.5 mm is used. When welding, hold the tin wire in your left hand and the soldering iron in your right hand. When the soldering iron contacts solder joints, the solder should be sent at the same time, and the amount of solder should be appropriate. Too much solder tends to cause a short circuit in lap welding, while too little solder cannot fasten components.

Do not move the soldering head back and forth on the solder joint and press it down hard. Increase the contact surface between the soldering iron and the solder joint, or increase the heat transfer area to speed up the welding speed and shorten the welding time. It should be noted that if the contact time between the soldering iron and the welding point is too short, and if the heat supply is insufficient, the solder surface of the joint is not smooth, the crystallization is rough and brittle, and then the welding spot is not strong enough to form incomplete welding and false welding. Lack of solder at the joint is unable to make components firm and may also burn out electronic components and printed circuit boards. Therefore, the amount of solder should be moderate, that is, the parts feet are all immersed in solder, and their outlines are vaguely visible. After the solder joint is welded, remove the soldering iron and wait for the solder to solidify for a moment. If the welding part is moved before solidification, the solder will solidify to be sand-like, resulting in weak adhesion and false welding. Solder joint contrast is shown in Figure 3-13.

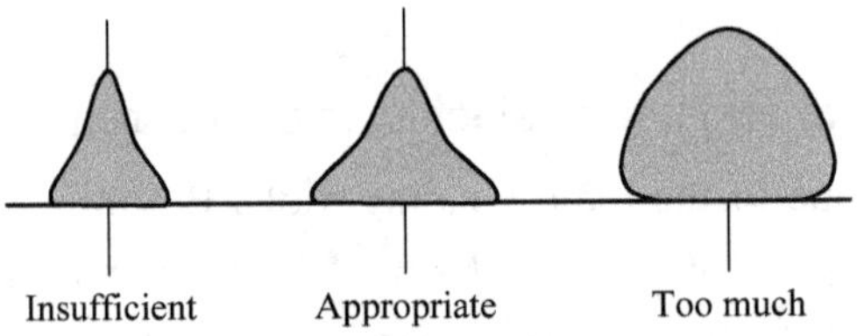

Fig. 3-13 Solder joint contrast diagram

After the welding is finished, first check whether there is any missing welding, lap welding, incomplete welding and other phenomena or not. Incomplete welding is a relatively difficult problem to find. There are many factors causing incomplete welding. When inspecting, each component can be gently pulled by pointed pliers or tweezers to see if it shakes. Rewelding should be carried out if any shaking is found.

【Practice and Operation】

1. Task preparations

(1) Familiarize with welding safety operation rules and precautions before training.

(2) Prepare electronic components such as soldering iron, solder, circuit board, wire, resistance and capacitance(as shown in Figure 3-14).

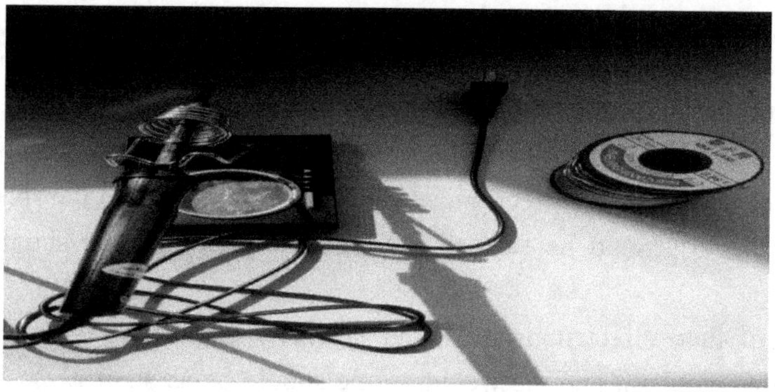

Fig. 3-14 Electric solderıng iron and soldering tin equipped in training room

2. Task contents

1) Preliminary testing of components with a multimeter

(1) Method for the testing of resistance.

The pointer of the multimeter touches the two pins of the resistance separately, not to measure it with the resistance in hand. Otherwise, the multimeter cannot measure the resistance accurately because of the interference of human resistance. Set the multimeter to the resistance gear and hit the range "×10k", "×100k" and "×1 k"respectively. The reading on the multimeter indicates that the resistance is in good condition.

(2) Quality inspection of capacitor.

The electric leakage of capacitors should be detected before they are used.

Capacitor of capacitance in 1-100 μF is measured by "R×1k" gear; capacitor of capacitance larger than 100 μF is measured by "R×10k" gear. Specific steps are as follows: the multimeter pens are connected to the ends of the capacitor separately, and the pointer should swing to the right first, and then return to the place where the resistance is infinite. Exchange the meter pens to test the capacitor. If the pointer is close to or at"∞", the leakage resistance is large and the capacitor is good in performance; if the pointer is far from the infinite point, the leakage resistance is small and the capacitor is poor in performance; if the pointer is always fixed at "0", a short circuit occurs in the capacitor.

(3) Quality inspection of inductor.

The inductance of an inductor is usually measured by inductance meter or multimeter. The DC resistance of an inductor is very small being of a few tens of an ohm. When the resistance of the coil measured by multimeter is infinite, the internal circuit is broken.

2) Practice of welding components and wires on PCB board

(1) Welding of resistor: it is required to weld resistors of one specification before welding resistors of another specification.

(2) Welding of capacitor: note that the polarity direction of polar capacitors cannot be wrong. The type mark on the capacitor should be easy to see and the direction should be consistent. The electrolytic capacitor must be firmly attached to the PCB board and cannot be

suspended.

(3) Welding of diode: after correctly identifying the positive and negative polarity, fix it into the specified position according to the requirements, the model mark should be easy to see, and the welding time should be as short as possible.

(4) Welding of triode: after correctly identifying each pin, fix it to the specified position as required, the type mark should be easy to see, and the welding time should be as short as possible.

(5) Welding of field effect transistor: after correctly identifying each pin, fix it into the specified position as required, and the welding time should be as short as possible. If heat sink needs to be added, polish the contact surface and add silicone grease before fastening.

(6) Welding of integrated circuit (chip): please check the model and welding position according to the requirements in the drawings. When welding, first weld the two pins on the edge of the chip to fix a position, and then weld point by point from left to right or from top to bottom. Welding time should be as short as possible. Avoid pull welding.

3. Task procedures

(1) Prepare the equipment needed shown in Table 3-1.

(2) Make a preliminary inspection of the quality of components.

(3) Practice welding, welding wire, resistor, capacitor, diode, transistor, field effect tube, and integrated circuit chip.

4. Precautions

(1) When welding, the amount of solder should be appropriate. If there is not adequate solder, it is not easy to weld, while it is easy to cause short circuit and affect the appearance if there is excess solder. The duration of welding should be determined according to the actual situation.

(2) After the completion of welding, check whether there is any missing welding, bump welding and false welding, and clean the solder at the welding spot.

(3) In the process of welding, please note that the electric soldering head should not be pointed at a person. When it is not in use, it should be inserted back into the support and not placed on the training platform at will. Take care to keep the sponge moist.

(4) After the work is done, first cut off the power, then let the soldering iron cool down, and place it to the designated position.

5. Notes and summary

____

Task 3 Manufacturing of DC Voltage Stabilizing Power Supply

【Relevant Knowledge】

Almost all electronic circuits need a stable DC power supply. When checking and repairing the indicator, besides the appropriate standard instrument, there must also be a suitable DC power supply and adjusting devices. When power is supplied by an AC network, it is necessary to convert the AC power supplied by the grid into a stable DC power. After rectification and filtering, AC becomes DC. Although it can be used as DC power supply, the output DC voltage will fluctuate with the fluctuation of grid voltage. At the same time, the load current in use is also constantly changing, some of which may vary greatly. When it flows through the internal resistance of the rectifier, it will produce a fluctuating voltage drop on the internal resistance, so that the output voltage will fluctuate with the fluctuation of the load current. If the load current is small, the output voltage is high; if the load current is large, the output voltage is low. The voltage fluctuation of DC power supply will cause the instability of circuit operation. For precise measuring instruments, automatic control or electronic calculating devices, it will cause errors in measurement and calculation,or even make these devices fail to work normally. Therefore, the DC voltage stabilizing power supply is usually needed.

The transistor DC regulated power supply can be used as DC power supply for various transistor instruments, electronic computers, automatic control systems and equipment. Precise stabilized voltage and current supply can also be used as stabilized voltage and current supply for the inspection of some electrical instruments. Therefore, the transistor DC voltage stabilizing power supply is a necessary instrument commonly used in scientific research, production, teaching and maintenance units.

The DC regulated power supply generally consists of four parts: transformer, rectifier, filter and regulator. The transformer reduces the voltage of 220 V AC power supply to the required low-voltage AC. The rectifier converts the low-voltage AC into DC. After that, the AC component in DC is removed by filtering circuit. Finally, the regulator converts the unstable DC voltage into stable DC voltage for output.

1. Principle of DC stabilized power supply

1) Block diagram and working principle

The block diagram for a typical circuit of the transistor series DC stabilized power supply is shown in Figure 3-15. It consists of rectifying filtering circuit, series voltage regulating circuit, auxiliary power supply and protection circuit.

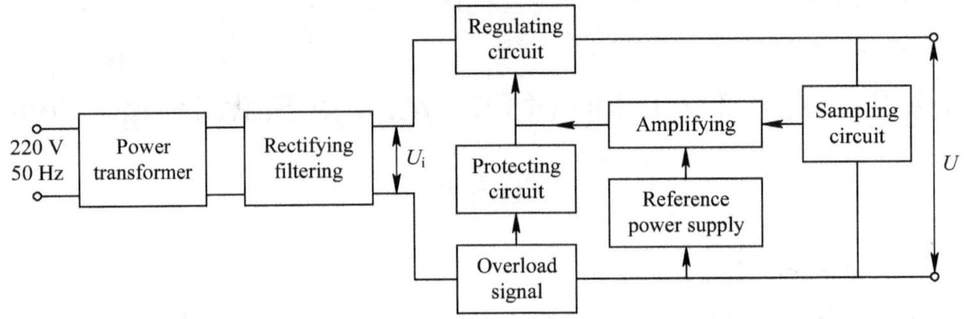

Fig. 3-15 Block diagram for DC stabilized power supply circuit

Rectifying filtering circuit includes power transformer, rectifying circuit and filtering circuit. The DC power supply commonly used in semiconductor circuits has rated voltage values of 6 V, 12 V, 18 V, 24 V and 30 V, while the grid voltage is generally 220 V. In order to convert the AC voltage of the grid into the required DC voltage, it is necessary to first step down the voltage across the power transformer, and then turn AC into pulsating DC through the rectifying circuit. It must be converted into pulsating DC because the voltage after rectification has a larger AC component. The DC voltage can be smoothed by filtering circuit.

Although the DC voltage obtained by the filtering circuit has little fluctuation, the value of the voltage is still unstable. There are three main reasons. First, the voltage of the AC network generally fluctuates by about±10%, which will cause the DC voltage of the rectifying filtering output to fluctuate by about ±10%. Second, the rectifying filtering circuit has internal resistance, and when the load current changes, it will fluctuate by about ±10%. The voltage drop on internal resistance will also change, so that the output DC voltage will also change. Third, in the rectifying voltage stabilizing circuit, the output voltage is unstable because the characteristics of the semiconductor devices used vary with the ambient temperature.

Voltage regulating circuit can maintain the stability of the output DC voltage, so that the output DC voltage does not change with the grid voltage, load or temperature. The series voltage stabilizing circuit is composed of regulating circuit, comparative amplifying circuit, sampling circuit and reference voltage. The regulating tube in the regulating circuit is connected in series between the filtering circuit and the load, so it is called the series regulator circuit. The regulator is equivalent to a variable resistor. If the output voltage increases, the resistance value increases correspondingly and the output voltage drops back. If the output voltage drops, the resistance decreases and the output voltage increases. The purpose of voltage stabilization can be achieved by adjusting the output voltage for it to remain unchanged.

Sampling circuit uses resistive voltage dividing method to sample the change of the output voltage in a certain proportion as sampling signal. The reference voltage is a stable and standard reference voltage. Sampling signal is compared with reference voltage which is applied to comparative amplifying circuit at the same time, and then their difference is amplified. The amplified voltage is used to control the base injection current of the regulator, thus changing the

DC internal resistance of the regulator and adjusting the output voltage for it to be stable. In order to improve the performance of voltage regulator, two-stage differential amplifier is often used in comparative amplifying circuit, which has larger amplification ratio and stronger control ability. Secondly, comparative amplifying circuit requires smaller zero drift and better temperature stability.

The above rectifying filtering circuit is combined with the series voltage stabilizing circuit, which is also known as the main power supply. The principle of voltage stabilization is as follows: if the output voltage increases due to the variation of the grid voltage or load, the sampling voltage generated by the sampling circuit also increases. At this time, the sampling voltage is larger than the reference voltage, and the voltage difference will be amplified by the comparative amplifying circuit, the regulating circuit reduces the emitter voltage of the regulator, its base current decreases, the DC internal resistance of the regulator increases, and its voltage drop increases, so that the output voltage is reduced, maintaining the stability of the output voltage. Similarly, when the output voltage is reduced, through a similar process, the DC internal resistance of the regulating tube is reduced, and the voltage drop of the tube is reduced, and the output voltage is also increased, thus basically remaining unchanged.

In addition to the main power supply, there are generally two sets of auxiliary power supply. The first auxiliary power supply consists of a rectifier and a voltage regulator, and its output voltage is quite stable. The second auxiliary power supply is similar to the main power supply circuit, and also consists of a rectifying filtering circuit and a series voltage regulating circuit. Its output voltage is very stable. On the one hand, the output voltage of the first auxiliary power supply serves as the power supply voltage of the protecting circuit. On the other hand, after being connected in series with the output voltage of the main power supply and the output voltage of the second auxiliary power supply, it serves as the power supply voltage of the last differential amplifier of the comparative amplifying circuit of the main power supply, which provides a voltage stabilized power supply of a higher voltage for the comparative amplifier circuit, so that its gain is larger. The regulating sensitivity of series voltage regulating circuit of the main power supply is improved, and the stability of its output voltage is further improved. On the one hand, the output voltage of the second auxiliary power supply is used as the power supply voltage of the differential amplifier of the comparative amplifying circuit of the main power supply. On the other hand, it is used as the reference voltage of the comparative amplifying circuit of the main power supply through the voltage dividing circuit.

When overload occurs in series voltage regulating circuit, especially in the case of short circuit at the output, almost all the input DC voltage falls on both ends of the regulator. Even for a short time this overload phenomenon will cause the regulator and rectifier diode to burn up immediately. Therefore, it is necessary to adopt a fast-acting over-current automatic protection circuit. When overload or short circuit occurs, the regulator is cut off by the protection circuit. At this time, the output voltage and current are basically reduced to zero, playing a protective role. This protection circuit is called cut-off protection circuit.

2) Series voltage stabilized circuit

Figure 3-16 shows a series transistor voltage stabilized circuit with an amplifier.

The input voltage U_i is supplied by the rectifying filtering circuit. Resistors R_1 and R_2 form a voltage divider, which applies a part of the variation of output voltage to the input of the amplifier composed of V_{T1}, so it is called sampling circuit. Resistor R_3 and regulator V_Z constitute regulator circuit to provide reference voltage, so that the emitter potential of V_{T1} is fixed. Transistor V_{T1} acts as an amplifier, which can compare and amplify signals. R_4 is the collector resistance of V_{T1}. The signal output from the collector of V_{T1} is directly applied to the base of the regulator V_{T2}.

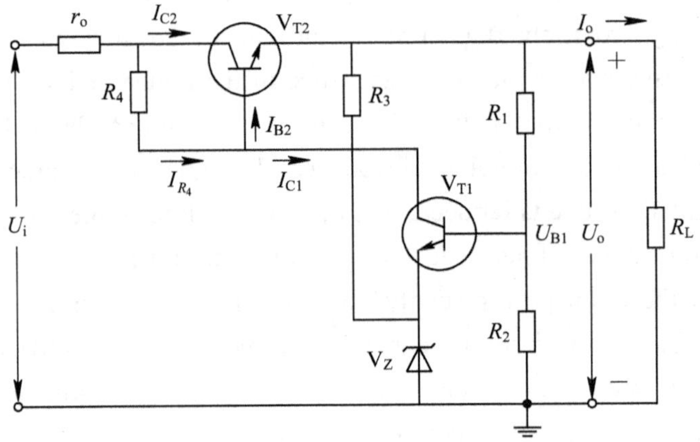

Fig. 3-16　Series voltage stabilizing circuit

If the output voltage U_o decreases due to the decrease of grid voltage or the increase of load current, the base potential U_{B1} of V_{T1} decreases through the voltage dividing effect of R_1 and R_2, and the emitter potential U_{E1} of V_{T1} is basically unchanged by the regulator V_Z. As a result of comparison, the forward voltage of the emitter junction of V_{T1} decreases, and thus the I_{C1} of V_{T1} decreases and the U_{C1} increases. The increase of U_{C1} increases I_{B2} and I_{C2} of V_{T2}, decreases U_{CE2}, and finally increases the output voltage U_o to close to the original value. The above voltage stabilization process can be expressed as:

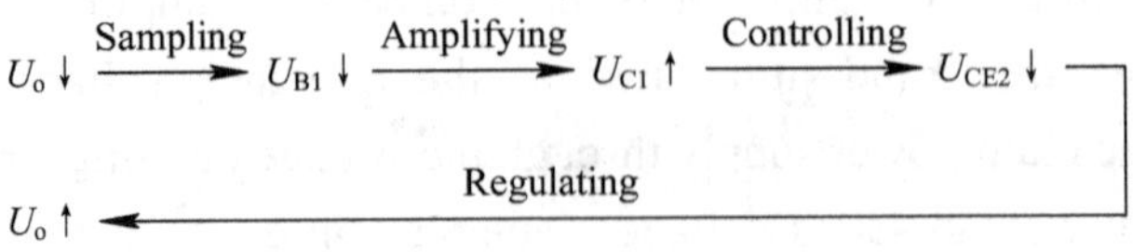

Similarly, when U_o rises, it remains basically unchanged through the stabilization process.

The comparative amplifier can be a single transistor amplifier circuit, but in order to improve its gain and temperature stability of the output voltage, it can also be multi-stage differential amplifier circuit and integrated operational amplifier. The regulator is usually a power transistor. The β value is increased, so that the small current of the comparative amplifier can drive the power transistor. It can also be a composite transistor composed of two or three transistors. If the power of the regulator can not meet the requirement, several regulators in parallel can be used to increase the branch circuit in order to expand the output current.

The method for the connection of sampling circuit is different because of different uses: for voltage stabilizer, the sampling resistance is parallel to the load; for current stabilizer, the sampling resistance is in series with the load.

Some electronic devices require two-way supply voltages of equal magnitude and opposite polarity. Such power supply voltage can be obtained through a symmetrical two-way voltage stabilizing circuit.

3) Auxiliary power supply circuit

(1) First auxiliary power supply circuit.

In the circuit shown in Figure 3-16, the load R_4 of the amplifier V_{T1} is directly connected to the variable input voltage U_i, so the change of the input voltage will directly act on the base of the regulator V_{T2} through R_4, thus changing the output voltage and affecting its stability. To overcome this shortcoming, an independent auxiliary power supply U_{Z2} can be used, as shown in Figure 3-17. This power supply, also known as the first auxiliary power supply, is a voltage stabilizing circuit composed of R and U_{Z2}. The voltage U_{i1} is obtained by rectifying and filtering another secondary winding of the same transformer. The voltage U_{Z2} is obtained by the voltage stabilizing circuit, which is connected in series with U_o as the power supply of V_{T1}. Because U_{Z2} and U_o are quite stable, the influence of the fluctuation of power supply voltage on the output voltage can be greatly reduced.

Because U_{Z2} and U_o are added as the power supply of comparative amplifier, R_4 can be selected larger than the original one to improve the amplification factor, thus further enhancing the control ability and improving the stability of output voltage.

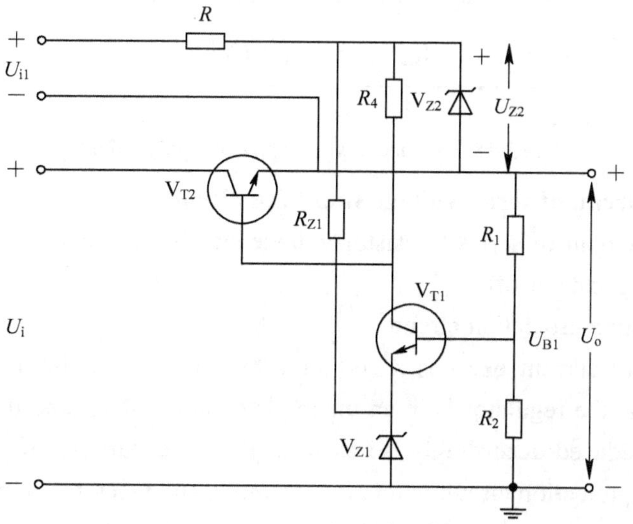

Fig. 3-17 First auxiliary power supply circuit

(2) Second auxiliary power supply circuit.

In the circuit shown in Figure 3-16, the output voltage U_o of the series regulator circuit can be expressed by the following formula:

$$U_o = (U_Z + U_{BE1})\frac{R_1 + R_2}{R_2}$$

It can be seen that the output voltage can be adjusted by changing the voltage dividing ratio of the sampling circuit. The smaller the R_1 is, the smaller the output voltage U_o is. When R_1=0, the output voltage is the lowest, and its value is $U_{omin}=U_Z+U_{BE1}$, that is, the lowest output voltage is still higher than the voltage U_Z of the regulator. The disadvantage of this circuit is that the output voltage can not be adjusted to zero. In order to expand the adjusting range of output voltage, a second auxiliary power supply can be added. As shown in Figure 3-18, the voltage of the regulator of this circuit is supplied by U_{i2} of another group of rectifier circuits. It can be seen from the figure that if R_1=0, $U=U_{BE1}\approx 0$. It can be seen that the second auxiliary power supply provides the possibility of adjusting the output voltage close to zero. As long as the divider ratio of the sampling circuit is changed, the output voltage can be continuously adjusted in a wide range.

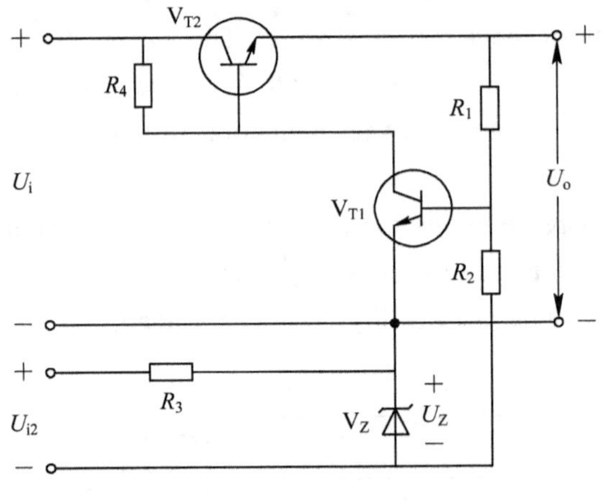

Fig. 3-18 Second auxiliary power supply circuit

4) Protection circuit of series voltage stabilized circuit

The protection circuit of series transistor voltage stabilized circuit can be divided into two types, current limiting and cut-off.

(1) Current limiting protection circuit.

When the output current exceeds a certain value, the current limiting protection circuit starts to work, so that the regulator is in an incomplete cut-off state, and the output current and output voltage are reduced accordingly, so as to achieve the purpose of protecting the power supply. This kind of protection circuit is relatively simple, and when the output overload or short circuit is eliminated, the voltage stabilized circuit will automatically resume work.

The dotted line enclosure shown in Figure 3-19 is a common current limiting protection circuit. V_{T3} is called protective tube. The output voltage is divided by R_5 and R_6, and the voltage on R_6 is taken to provide reverse bias for V_{T3} base. R_7 is the detection resistance, and its resistance value is small. The voltage drop of the output current on R_7 provides a forward bias to the V_{T3} base.

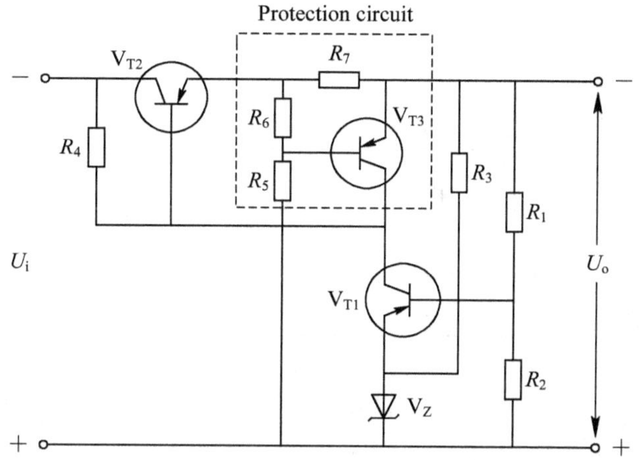

Fig. 3-19 Current limiting protection circuit

Under normal circumstances, the reverse bias on R_6 exceeds the forward bias on R_7, so V_{T3} is in the cut-off state and has no effect on the operation of the voltage stabilizing circuit.

When the output current is too large due to overload, the forward voltage drop of R_7 increases, and V_{T3} is turned on. Thus, the voltage at both ends of V_{T3} tube decreases, and the forward voltage of the emitter of the regulator tube V_{T2} decreases, so that the current of regulator tube decreases, and the output current and voltage decrease, which plays a protective role for regulator tube.

The necessary condition for the protection circuit to maintain V_{T3} on is that the output current generates forward bias through R_7, so the output current can only be reduced to a certain extent, but the regulator cannot be cut off. When the cause of output overload is eliminated, it can automatically return to normal state. The advantage of the protection circuit is simplicity and reliability, but the disadvantage is that the regulator still consumes a large amount of power when overload occurs.

(2) Cut-off protection circuit.

Cut-off protection circuit refers to a condition that when overload or short circuit occurs, the regulator is cut off by the protection circuit. At this time, the output voltage and current are basically reduced to zero, thus playing a protective role. The cut-off protection circuit is slightly more complex. It can be divided into two situations: one is that it can automatically resume work; the other is that after troubleshooting, the reset button is pressed or the AC power supply is cut off to restart, so that the regulated power supply can resume normal work.

The dotted line enclosed part shown in Figure 3-20 is the cut-off protection circuit.

Resistance R_8, voltage regulator V_{Z2} and voltage divider R_4 and R_5 provide base voltage for the protection tube V_{T3}. Output voltage U_o is supplied to V_{T3} emitter voltage through voltage divided by resistance R_6 and R_7. Detection resistance R is connected between R_7 and R_5. Output current I_o flows through R, resulting in voltage drop. The polarity of voltage across R_5, R_7 and R is shown in the figure. The emitter voltage applied to the protection tube V_{T3} is

$$U_{BE3} = U_{R_5} + U_R - U_{R_7}$$

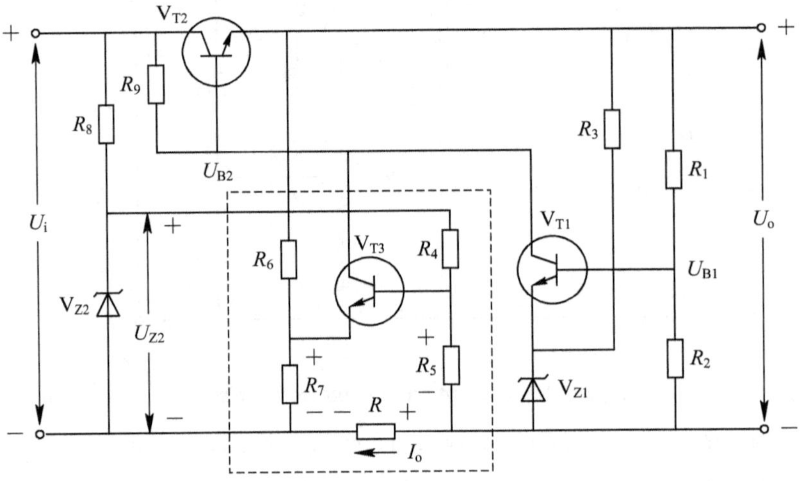

Fig. 3-20 Cut-off protection circuit

When the voltage regulating circuit works normally, I_o is within the rated value, $U_R = I_oR$, so that $U_{R_5} + U_R < U_{R_7}$, then U_{BE3} is negative, and V_{T3} emitter is reversely biased and reliably cut off. The protection circuit does not work and has no effect on the normal operation of the voltage stabilizing circuit.

When the output current I_o exceeds the rating value, the voltage across R increases, so V_{T3} is turned on. That is to say, the U_{B2} of the regulator V_{T2} decreases, which leads to the cut-off, the increase of U_{CE2} and the decrease of output voltage U_o. As a result, the voltage U_{R_7} across R_7 decreases, which further leads to the turn-on of V_{T3} and the formation of a positive feedback process, so that the regulator V_{T2} is cut off rapidly, and the output voltage and the output current are close to zero. At this time, the voltage U_{R_5} on R_5 is used to maintain V_{T3} on and V_{T2} off, which achieves the purpose of protection.

2. Use of DC voltage stabilizing power supply

The use of DC voltage stabilizing power supply is very simple. When using it, attention should be paid to the polarity of the required DC voltage. If positive voltage output is needed, the output end of the DC regulated power supply should be connected to the ground end of the electrical equipment, and the positive terminal should be connected to the required positive voltage end "+". If negative voltage output is needed, the above wiring should be reversed. Before you turn on the power, use a multimeter to measure the output voltage to check whether it meets the requirements, so as to avoid the damage of electrical equipment caused by excessive voltage.

In order for the electrical equipment to work properly and not to affect the stability and reliability of the electrical equipment due to the poor performance of DC power supply, it is better to test it before it works as a voltage stabilizer. The main contents of the test are adjusting range of output voltage, stability degree, ripple voltage and over-current protection, etc..

3. Maintenance of DC stabilizing power supply

1) Maintenance procedures

(1) Preliminary surface inspection.

Various voltage regulators are generally equipped with fuse wires for overload or short-circuit protection and input and output terminals. First it should be checked whether the fuse wires are fused or loosened, whether the terminals are loosened or short-circuited to the ground, and whether the meter pins of the voltage indicator are blocked. Then open the cover plate of the chassis to see if the power transformer is scorched or mildewed, and whether the resistance and capacitance are scorched, mildewed, cracked and other obvious damage signs.

(2) Measurement of rectifying output voltage.

There are one or more rectifying output voltages in all kinds of regulated power supply. If one of these rectifying output voltages is abnormal, the regulated power supply will have various faults. Therefore, when overhauling, it is necessary to measure whether the rectifying output voltage is normal in the first place.

(3) Testing of electronic devices.

If the rectifying output voltage is normal and the output voltage is not normal, it is necessary to further test whether the performance of the regulator, amplifier tube is good, and whether the capacitor has the problem of breaking through short circuit or open circuit. If damaged or variable devices are found and replaced, the regulator power supply can normally be restored to normal.

(4) Check of the operating point of the circuit.

If the output of rectifier voltage and related electronic devices are normal, the operating point of the circuit should be further checked. For a transistor, there must be a certain working voltage between its collector and emitter, and a bias voltage between the base and emitter. The polarity of the transistor should meet the requirement and it should work in the amplification region.

(5) Analysis of circuit principles.

There are two possibilities if voltage of the operating point of a transistor is found to be abnormal. One is that the transistor is damaged; the other is that other components in the circuit are damaged. At this time, we must carefully analyze the causes of the problems according to the circuit schematic diagram, and further identify the damaged and variable components.

2) Case study of common faults of the maintenance of regulating power supply

(1) Voltage regulation without stabilization.

When using a regulating power supply, it is usually to turn on and preheat the power first, then adjust the output voltage by "roughly" adjusting potentiometer, and observe whether the voltage regulation effect and regulation range are normal, and finally obtain the required power supply voltage value and connect the load. The voltage is normal when there is no load, but the output voltage drops after the load is connected. If the possibility of an external circuit fault is excluded, the fault at this time is that there is not voltage stabilizing effect by the regulating power supply.

The multimeter can be used to measure the on-off condition between the collector and the emitter of the high-power regulating tube. If a problem is not found, you can further check whether the rectifier diode is damaged. As long as there is rectifier tube damage, the full-wave rectifier will become a half-wave rectifier. When there is no load, the filter capacitor of large capacity can still provide enough rectifier output voltage to ensure the voltage regulation function of the stabilized output. After the load is connected, the output voltage of the rectifier will immediately drop, and the voltage of the stabilized output will also drop, thus losing the function of voltage regulation.

(2) Excessive output voltage, no voltage regulation or stabilization.

The output voltage of transistor DC regulated power supply exceeds the prescribed value in no-load condition, and there is no regulating or stabilizing effect. There are the following two faults.

① The collector and emitter of one of the composite regulator break down the short circuit, and the output voltage of the rectifier is directly applied to the output of the voltage regulator through the short circuit transistor, which is not controlled by the voltage regulation and stabilization.

② The collector or emitter of the sample amplifier is interrupted, and the composite regulator is directly under the negative voltage of the auxiliary power supply. The base current is very large, which makes the internal resistance between the emitter and collector of the regulator become very small, and the rectifier output voltage is directly applied to the output voltage of the regulator.

(3) The output voltage of each stage is very small and shows no voltage regulation effect. Faults may occur in the following situations:

① No rectifier voltage output comes from the main rectifier.

② The voltage of auxiliary power supply is zero, causing the regulating tube not to work.

③ The sampled amplification tube's collector-emitter reversely breaks down short circuit, causing the regulating tube not to work.

【Practice and Operation】

1. Task preparations

(1) Preview the working principle of DC regulated power supply and other welding cases.

(2) Prepare the specific components and welding tools needed for DC voltage stabilized power supply:

One three-terminal voltage regulator 7805, 78 series being positive output, output being +5 V;

One electrolytic capacitor of 1000 μF/16 V and one 470 F/16 V for filtering;

Two porcelain dielectric capacitors of 0.1 μF , with low frequency of high frequency resistance for filtering;

Four rectifier diodes 1N4007 to construct bridge rectifier circuit, to achieve the rectifier function;

One resistor of 470 Ω;

One LED light emitting diode;

One binding post;

Universal plate.

2. Task contents

The components are welded and tested according to the circuit schematic diagram(as shown in Figure 3-21) of DC stabilized power supply.

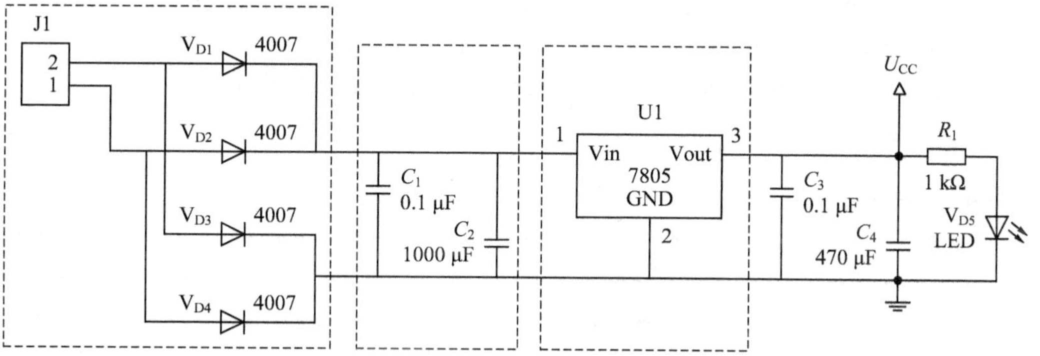

Fig. 3-21 Schematic diagram of DC stabilizing power supply

3. Task Procedures

(1) Testing the function of electronic components.

(2) Welding of the circuit.

Apply a layer of flux to the circuit board that has been made. Install the components on the circuit board according to the schematic diagram and check whether the components are in the correct position. After the inspection, each component is soldered firmly with solder by electric soldering iron to ensure that there is no false welding for each component. When welding components, the welding time is minimized according to the heat resistance of different components. After welding, check the circuit breaking and short circuit with a multimeter.

(3) Debugging of the circuit.

Apply 220 V AC voltage; check whether the circuit is abnormal. Then check whether the output voltage is normal with the multimeter. Connect the AC millivolt-meter to observe its AC ripple voltage, and the circuit is normal if the ripple voltage is small.

(4) Testing of the circuit.

The output voltage is +5 V measured by a multimeter (the actual measured voltage is slightly less than +5 V, because of the loss of part of the circuit voltage).

(5) Analysis of measurement results.

The RMS is measured by AC millivolt-meter, but because the ripple is not sinusoidal, there are some errors. The wires in the circuit board will lead to certain errors; there are also some errors in reading. All the measurement results within the error range are normal.

4. Precautions

(1) Components should not be wrongly fixed, left out, wrongly connected, tilted or loosely fixed.

(2) Solder joints should be full of tin, without burrs, pinholes, bubbles, cracks, tin hanging, pull points, missed soldering, touch soldering, false soldering and other defects.

(3) No rust and other impurities should be found on the surface of the metal parts on the printed circuit board after welding.

(4) No spots, cracks, bubbles, whitening and other signs are allowed on the welded circuit boards. Copper foil and coatings should not fall off, not be wrapped, or layered.

(5) The pin or lead surface of components should be tin-coated evenly.

5. Notes and summary

____

____

____

____

【Extension Exercises】

Exercise 1 Making DC Digital Multimeter

1. Task objectives

(1) Make electronic products through the integrated digital multimeter welding assembly module to master the welding process and related quality inspection technology.

(2) Understand the basic working principle of DC digital multimeter.

2. Equipment

The digital multimeter installation and adjustment module provided by Yalong YL-135 training platform.

3. Task contents

(1) Understand the basic working principle of DC digital multimeter. The basic working

principle of DC digital multimeter is shown in Figure 3-22. Because the purpose of this exercise is for the learner to be able to make DC digital multimeter according to the drawing and to master the welding process, the principle will not be explained any more. The reader interested in the principle can learn from other resources.

(2) Carry out welding production practice according to Figure 3-23.

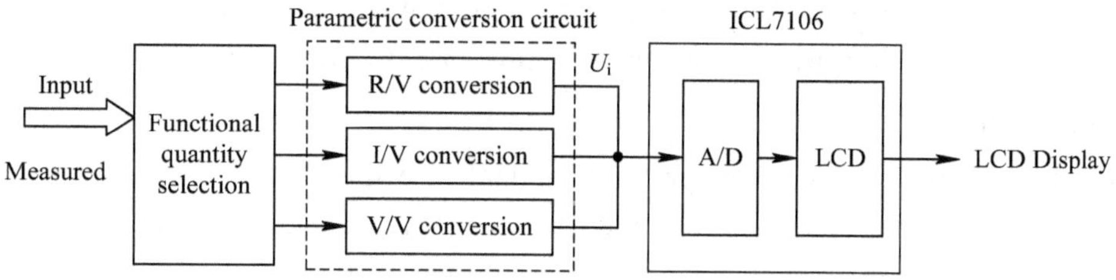

Fig. 3-22 Working principle diagram for DC digital multimeter

4. Task procedures

(1) According to Figure 3-23, check whether the equipment of the exercise is all in readiness and whether the model are right.

(2) Make the circuit of multimeter according to Figure 3-23. Weld the components in the kit on the PCB according to Figure 3-23.

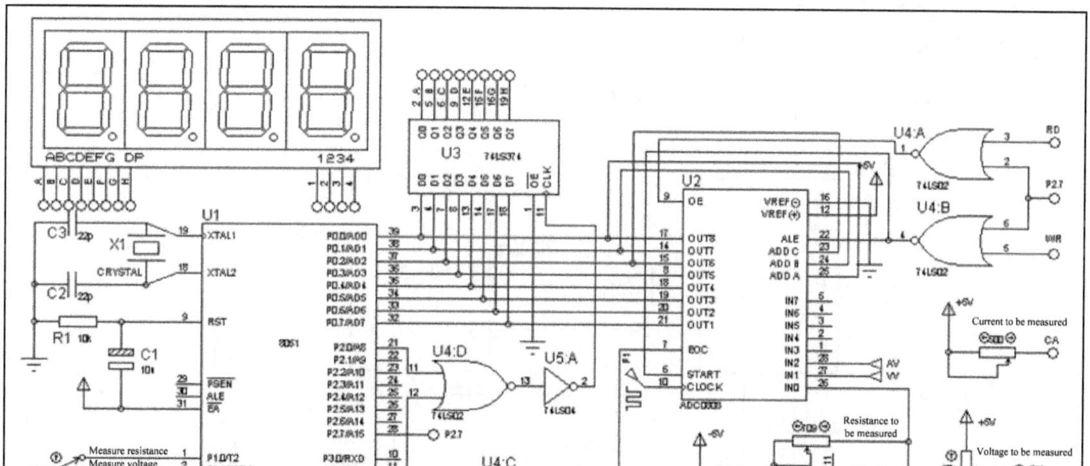

Fig. 3-23 Wiring diagram for DC digital multimeter

Exercise 2 Production of Dual-Tone Alarm

1. Task objectives

(1) Understand the basic working principle of dual-tone alarm.

(2) Through the production of alarm be further familiar with electronic products production process.

2. Equipment

The electrical components required for the production of this electronic product are as follows:

three 10 kΩ resistors; one 110 kΩ resistor; two 10 μF micro-capacitors; two 103 capacitors; one 200 kΩ variable resistor; two 555 integrated chips; one switch; one buzzer.

3. Task contents

(1) Understand the basic working principle of dual-tone alarm.

The principle of the dual-tone alarm is shown in Figure 3-24. When IC1 555 forms a self-excited multi-vibrator, its 5 pins are grounded through a small capacitance to prevent the influence of external interference on the threshold voltage. When it needs to be turned into a controllable multi-vibrator, a controllable voltage can be applied to the 5 pins of the circuit. This voltage will change the comparison level in the chip, thus changing the oscillation frequency. When the control voltage increases, the oscillation frequency will decrease, which is the modulation of the control voltage over the oscillation frequency. By using the modulation method, a dual-tone alarm circuit can be formed.

(2) Welding of the circuit.

(3) Debugging of the circuit.

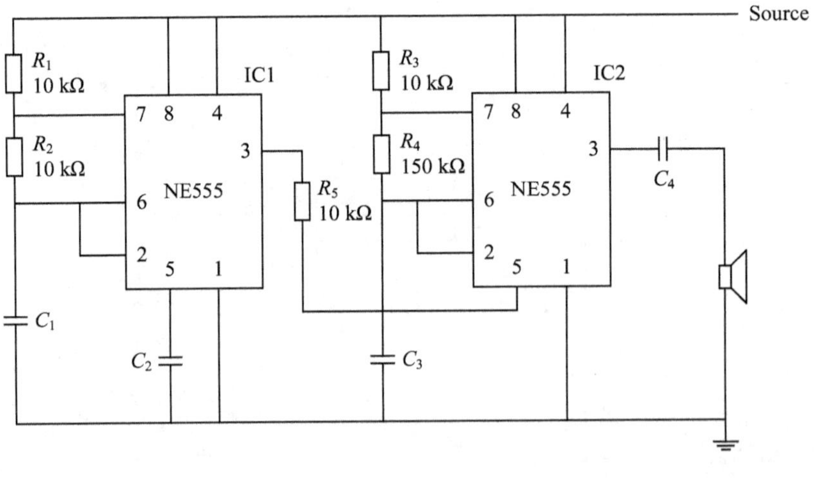

Fig. 3-24　Principle diagram of dual-tone alarm

4. Task procedures

(1) Welding.

When welding the circuit, first directly weld the chip on the universal board, then plug in the components according to the PCB layout, and next fix them. In principle, the lower components should be welded first. Therefore, the higher components, such as 10 μF micro-capacitors, buzzer, switch, etc., are not welded first, and then welded in proper order to complete the whole circuit.

(2) Debugging.

After the circuit board is welded and the PCB drawing is checked to confirm the circuit is correct, use a multimeter to test the circuit board for short circuit or missing welding. After the test, improve false welding and missing welding, then test again, and finally confirm that the circuit is OK. With power on, the buzzer will normally send out sounds of two frequencies, and the circuit debugging is completed.

Module Four Comprehensive Training in Digital Electronic Technology

【General Introduction】

This module designs three typical tasks to realize the functions of digital electronic circuits for the practitioners in the elementary electrician profession. The main task is to master the functional principles of basic digital chips and grasp the methods for the testing of the functions of different electronic chips. The contents include the functional test of basic logic gate circuit, the functional test of seven-segment code latch/decode/driver, and the functional test of two-way shift register. After this module, the trainees can meet the requirements of basic skills of digital electronic circuits for elementary electricians.

【Basic Elements】

1. Knowledge objectives

(1) Master the types and logical functions of basic gate circuits.

(2) Be familiar with the logical functions of NAND gates, NOR gates, AND-OR-NOT gates, EXO gates, and OD gates.

(3) Master the method for the pin arrangement and operation of the common cathode light emitting diode and digital tube.

(4) Understand the pin functions of COMS 4513 seven-segment code latch, decode and driver, and its limited full arrangement and the method for its use.

(5) Master the function and working principle of each pin of the bidirectional mobile register.

2. Competence objectives

(1) Be able to understand the logic function table of digital circuit device.

(2) Be able to measure the output voltage of CMOS and TTL logic gates at high and low levels.

(3) Be able to measure the influence of suspension of CMOS logic gate input on output.

(4) Be able to connect CD4513 and seven-segment LED digital tube correctly and apply them.

(5) Be able to apply pulse sequence generator made of bidirectional displacement register.

3. Operation standards

1) Preparation before the training

(1) Read the experimental instructions carefully, and make notes according to the relevant

contents of the relevant teaching materials. Finish the preview work.

(2) Clarify the experimental purpose, list the experimental items and procedures, and go to the laboratory to understand the structural characteristics of the experimental devices, experimental units and required accessories and instruments.

(3) Be familiar with and bear in mind the experimental precautions, and avoid personal safety accidents and equipment damage.

2) Task procedures

(1) Establishment of teams and division of labor.

The training task is carried out in a group of two to three people. The task usually includes choosing functional chips, building digital circuits, testing digital circuits and building complex functional circuits. The members of the experimental group should have a clear and reasonable division of labor, take their places, coordinate their work, and consult constantly in the project to ensure that the training operation is standardized and the results are accurate.

(2) Building digital circuits and testing.

Different digital circuits are built and their functions are tested according to the requirements of the training project.

After the data test is completed, the correctness of the data is checked according to the working principle of the project circuit. If it is different from the expected results, the quality of circuit connection or chip should be checked.

(3) Disconnecting and cleaning up.

The power supply must be switched off first after the task is completed, the wires, instruments and circuit unit modules should be sorted out, and the training room should be cleaned.

3) Training report

At the end of each program, students are required to write a training report. The report should highlight the process and contents of the task, record the measurement data or describe the function of the electrical chip. If problems are found in the process of practical operation, they should be timely recorded in the report for analysis and solution. The writing of a practical training report is not only the record of practical training operation process, but also an important link of learning, discovering and solving problems. It can comprehensively improve students' logical ability, graphic and text processing and data analysis ability. The report requires clear thinking, concise writing, neat charts and graphs, reasonable analysis, clear conclusions, standard and smooth writing.

4. Safety standards

(1) Guarantee reliable grounding protection for training equipment. Ensure safety measures of leakage protection, overload protection, insulated desktop, carpet and safety wiring.

(2) Before turning on the main switch of the test-bed, check whether the power supply of the test-bed +12 V, −12 V, +24 V, and −24 V is turned off.

(3) When doing the experiment, the total power supply should be turned off first, and then the low-voltage power supply should be turned on after the connection is made, and the correct connection of the power supply voltage should be ensured.

(4) When doing digital circuit experiments, try not to connect other pins directly with the power supply except the power supply pins of the chip. If the high-level voltage is applied, the appropriate resistance should be connected in series before the application of the high-level voltage.

(5) After the check of the wires by the instructor, and with the consent of all the members of the group, the system can be operated by electricity.

(6) When wiring and disconnecting, the power supply must be cut off.

5. Quality standards

(1) The "red" and "black" connecting wires should be used for power supply and ground wire connection, and connecting wires of appropriate length should be selected according to requirements to ensure neat and concise wiring of the circuit.

(2) For digital circuit waveform adjustment, the appropriate unit and position on the oscilloscope should be chosen, to facilitate the reading and comparison of waveform parameters.

【Training Requirements】

1. Preparation of tools, instruments and equipment

The tools and equipment listed in Table 4-1 should be prepared prior to the training.

Table 4-1　Required equipment for digital circuit

No.	Name	Model	Unit	Quantity
1	Multimeter		Piece	1
2	Dual in-line modules	IC2×2, BX07 light-emitting diode, R10 module, resistance of 300 Ω		
3	Module circuit for YL-135 training platform	CD4001, CD4011, CD4070, 74LS00	Piece	One for each
4	DC regulating power supply		Set	1
5	0-1 single-pole double-throw switch module	AX21	Piece	1
6	LED status display module	AX26	Piece	1
7	Single pulse signal generator	AX22	Piece	1
8	Clock pulse signal source	AX25，IC3，IC2	Piece	3
9	Common cathode nixie tube module with current limiting resistor	DPY1	Piece	1
10	Dual in-line 18 pins socket module	IC4	Piece	1
11	Different types of registers, encoders, decoders, selectors, total adders, comparators and other integrated circuit chips	CD4513, CD40194, 74LS196, CD4532, 74HC138, 74HC151, 74LS283, CD4585	Piece	One for each

2. Knowledge assessment points

(1) Master the logic function of basic gate circuit.

(2) Master the meanings and functions of common digital circuit chip pins.

(3) Master the basic functions of encoder, decoder and bidirectional shift register.

3. Skills assessment points

(1) Understand the logic function table of digital circuit devices.

(2) Distinguish the pins of commonly used NAND gates and other integrated chips.

(3) Determine the output voltage of logic gate circuits of CMOS and TTL series, and be skilled in reading waveform signal information with an oscilloscope.

(4) Be able to connect CD4513 with the seven-segment code LED digital tube and apply the bidirectional shift register to form the pulse sequence generator.

【Work Tasks】

Task 1 Functional Test of Basic Logic Gate Circuit

【Relevant Knowledge】

1. Common numerical system

(1) Decimal system: Decimal system numbers have 10 digits in total, 0, 1, 2, 3, 4, 5, 6, 7, 8, and 9. For decimal numbers to be added, they follow the rule of "every decimal is one" and for them to be subtracted, they follow the rule of "borrowing one is ten". In practical digital circuits, it is very inconvenient to adopt decimal system, because decimal system has ten digits. In order to separate them strictly, ten different circuit states must correspond to them, which is difficult to realize technically. Therefore, decimal digits are not usually used directly in practical digital circuits.

(2) Binary system: Binary numbers have only two different digits, 0 and 1. The carry rule is "two in one" and the borrow rule is "one in two". Binary system is the most widely used digital system in digital circuits. Because the binary numbers have only 0 and 1 digits, which is easy to represent by the state of the circuit or device. Also, the binary operation rules are simple.

(3) Hexadecimal system: The hexadecimal system has the carry rule of "one in every sixteen".

2. Conversion between numerical systems

(1) Conversion of binary numbers to decimal numbers. The conversion method is to write the binary weight expansion, and then add up the values in decimal system to get the equivalent decimal number.

(2) Conversion of decimal integers to binary numbers. The conversion method is to divide decimal integers by 2 to get the remainder one by one until the quotient is zero. The first residue

is the lowest number of binary digits. This conversion method is usually called short division and residual countdown method.

(3) Conversion of binary numbers to hexadecimal numbers. The conversion method is to divide the binary number from right to left, every 4 bits into a group, for the remaining bits less than 4, supplement the bits with 0 in the high-order digit, and then write the equivalent hexadecimal number for each group.

(4) Conversion of hexadecimal numbers to binary numbers. The conversion method is to replace each hexadecimal digit with the corresponding four-digit binary number.

3. The coding

(1) Binary-decimal code: A common binary-decimal code is a 4-bit binary number representation of a 1-bit decimal number, also known as BCD code.

(2) Character code: Also known as a character set code, is the encoding of characters in a character set into an object in a specified set (e.g., bit pattern, sequences of natural numbers, 8-bit groups, or electrical impulses) for the storage of text in a computer and its transmission over a communications network. A common example is the encoding of the Latin alphabet into Morse code and ASCII. In ASCII letters, numbers, and other symbols are numbered and this integer is represented by 7-bit binary. An additional expanded bit is often used so that it can be stored as 1 byte.

4. Introduction to basic logic gate circuit function

1) AND gate

The logical expression for AND gate is $Y=AB$. When both the input terminals A and B are 1, the output terminal Y is 1, or otherwise Y is 0, as shown in Figure 4-1. Common chip models for the gate are 74LS08, 74LS09, etc..

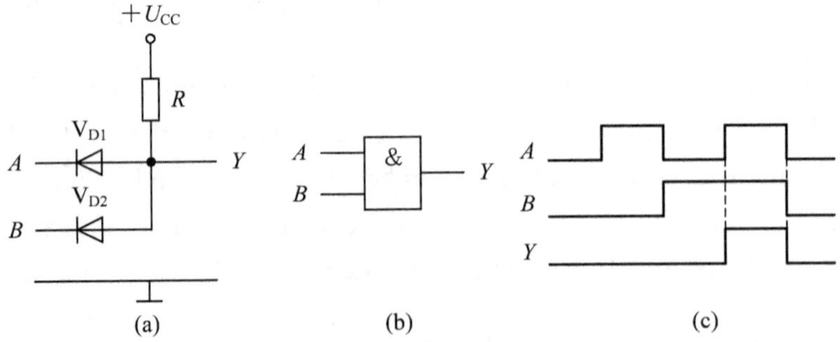

Fig. 4-1 AND gate logic circuit and input and output wave forms

2) OR gate

The logical expression for OR gate is $Y=A+B$. When one of the input terminals A and B is 1, the output terminal Y is 1, and when both of the input terminals A and B are 0, Y is 0, as shown in Figure 4-2. Common chip modules for OR gate are 74LS32, etc..

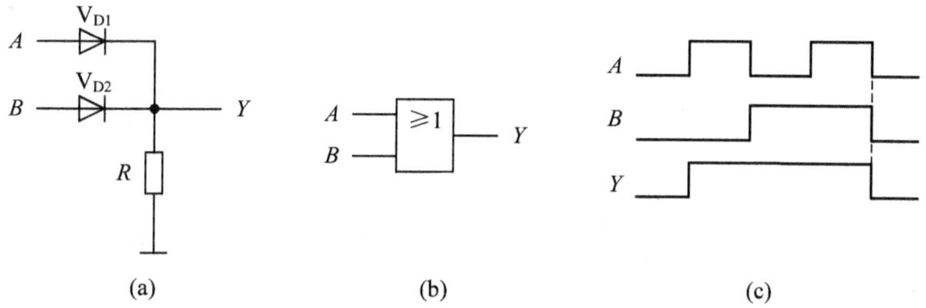

Fig. 4-2 OR gate logic circuit and input and output wave forms

3) NOT gate

The logical expression for NOT gate is $Y=\overline{A}$. The output is always the opposite of the input, as shown in Figure 4-3. Commonly used chip models for NOT gate are 74 LS04, 74 LS05, 74 LS06, 74LS14, etc..

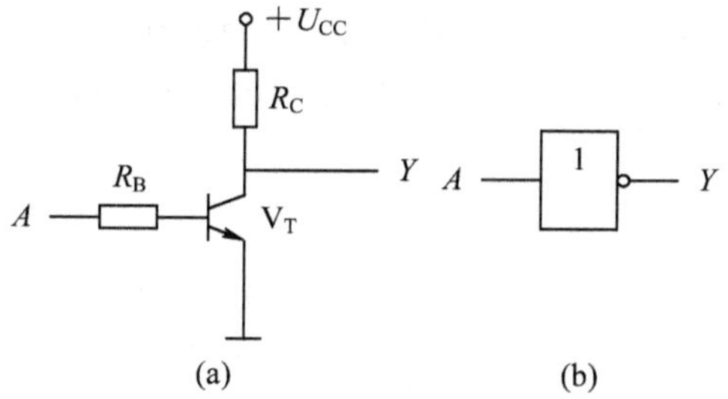

Fig. 4-3 NOT gate logic circuit

4) NAND gate

The logical expression for NAND gate is $Y=\overline{AB}$. Only when both the input terminals A and B are 1, is the output Y 0, or otherwise Y is 1. The logic circuit of NAND gate is shown in Figure 4-4. Commonly used chip models for NAND gate are 74LS00, 74LS03, 74LS31, 74LS132, etc..

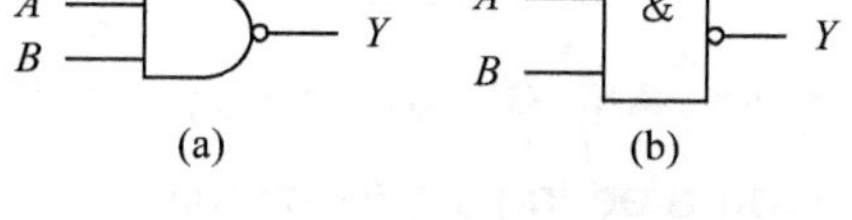

Fig. 4-4 NAND gate logic circuit

5) NOR gate

The logical expression for NOR gate is $Y=\overline{A+B}$. As long as one of the inputs A and B is 1, the output Y is 0. When the input terminals A and B are both 0, Y is 1. Its logic circuit is shown in Figure 4-5, and the common chip models for NOR gate are 74LS02, etc..

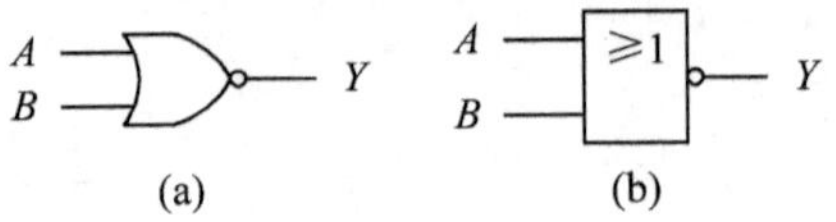

Fig. 4-5　NOR gate logic circuit

6) XNOR gate

The logical expression for XNOR gate is $Y = A \odot B$. If the inputs A and B are the same, the output Y is 0. If A and B are not the same, the output Y is 1. Its logic circuit is shown in Figure 4-6.

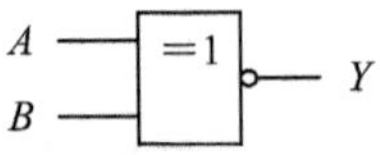

Fig. 4-6　XNOR gate logic circuit

7) Exclusive-OR gate

The logical expression for exclusive-OR gate is $Y = A \oplus B$. If inputs A and B are the same, the output Y is 1. The output Y is 0 if A and B are not the same. Its logic circuit is shown in Figure 4-7.

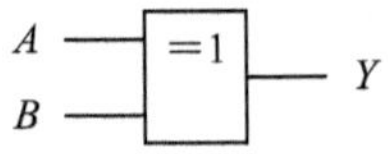

Fig. 4-7　Exclusive–OR gate logic circuit

8) AND-OR-NOT gate

The logical expression for AND-OR-NOT gate is $Y = \overline{AB + CD}$. The logical order of the four inputs A, B, C and D is first AND two pairs of them, then OR them and finally NOT. The logic circuit of AND-OR-NOT gate is shown in Figure 4-8.

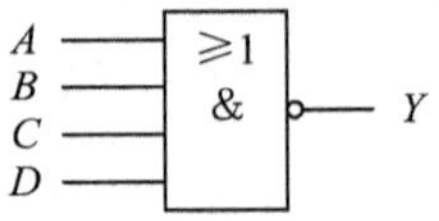

Fig. 4-8　AND-OR-NOT gate logic circuit

5. Basic introduction to integrated logic gate circuit

There are mainly two types of integrated logic gates, CMOS and TTL series. When working in positive logic state, high level is defined as logical state 1 and low level as logical state 0. If the input is +5 V, the state is 1 and if the input is 0 V, the state is 0. The voltage values of high and low output levels of these two series are different. If they adopt the +5 V power supply voltage, the voltages of the logical states 1 and 0 in CMOS series are close to +5 V and 0 V, while the voltages of the logical states 1 and 0 in TTL series are close to 4.3 V and 0.2 V.

For logical input variables and output variables, their values are only 1 or 0, and the basic logical operations are AND, OR, and NOT. The logic operations of AND are $1 \cdot 1=1$, $1 \cdot 0=0$, and $0 \cdot 0=0$; the logics of OR are $1+1=1$, $1+0=1$, and $0+0=0$; the logics of NOT are $\overline{1}=0$, $\overline{0}=1$; NAND gate is the combination of AND gate and NOT gate; NOR gate is the combination of OR gate and NOT gate. The logical operations of Exclusive-OR gate are $1 \oplus 0=1$, but when $1 \oplus 1$ and $0 \oplus 0$, the output is 0. NAND-OD gate drivers of CMOS are open drain circuit NAND gate, and its output terminal should be connected with an external load.

【Practice and Operation】

1. Task preparations

(1) Previcw the contents of the training project and the working principle of basic logic gate circuit.

(2) Prepare training instruments: multimeter, 0-1 single-pole double-throw switch module AX21, regulated power supply, light-emitting diode and driving circuit module AX26, dual-in-line module IC2 and BX07, light-emitting diode, R10 module, and resistance of 300 Ω. Training device modules: CD4001, CD4011, CD4070, and 74LS00.

2. Task contents

(1) Master and understand the basic logic gate circuit working principle and input and output state.

(2) Plug the training device into the dual-in-line module socket, and build the following logic test circuit as shown in Figure4-9.

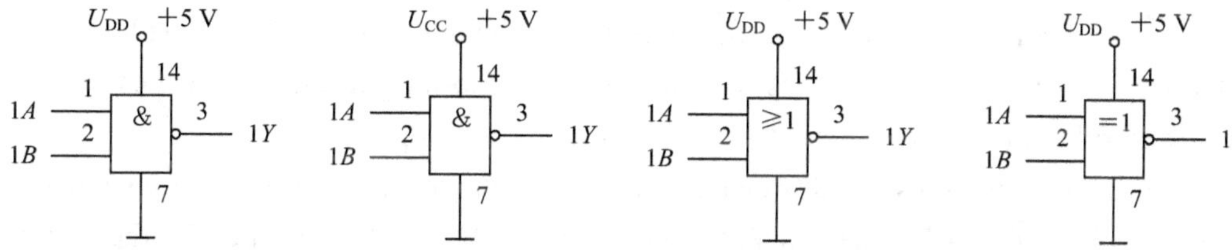

(a) 1/4 CD4011 NAND Gate (b) 1/4 74LS00 NAND Gate (c) 1/4 CD4001 NOR gate (d) 1/4 CD4070 Exclusive-OR gate

Fig. 4-9 Logic gate circuit to be tested

(3) Test the output logic state (with AX21 module) according to the input logic state (with AX26 module) in Tables 4-2 to 4-5.

Table 4-2 CD4011-2 input NAND gate

Input		Output	
1A	1B	1Y	Level/V
0	0		
0	1		
1	0		
1	1		
1	Suspended		

Table 4-3 74LS00-2 input NAND gate

Input		Output	
1A	1B	1Y	Level/V
0	0		
0	1		
1	0		
1	1		
1	Suspended		

Table 4-4 CD4001-2 input NOR gate

Input		Output	
$1A$	$1B$	$1Y$	Level/V
0	0		
0	1		
1	0		
1	1		
1	Suspended		

Table 4-5 CD4070 Exclusive-OR gate

Input		Output
$1A$	$1B$	$1Y$
0	0	
0	1	
1	0	
1	1	

3. Task procedures

(1) Insert the training device on the double-row in-line module socket. Consult the reference model in the appendix for the pin arrangement of the device.

(2) After adjusting the DC regulated power supply to +5 V, turn off the regulated power switch and connect it to the power supply of the training device and the pin socket of the ground terminal.

(3) Connect modules AX21 and AX26 to the power supply of +5 V.

(4) Connect the output of AX21 to the input of the device, while the output of the device is connected to any input hole of AX26.

(5) Turn on the regulator power switch, set 0 (0 V) and 1 (+5 V) on the AX21 0-1 switch according to the tables to observe the light-emitting diodes connected to the output terminal, 1 showing that the light is on, 0 showing that the light is off. The high and low voltage values of the output terminals of CD4011 and 74LS00 are tested by the multimeter DC voltage level.

(6) Test the input and output states of other devices according to the input states in Tables 4-2 to 4-5, and fill out the tables.

(7) Analyze the relationship between input and output logic function of each device in Tables 4-2 to 4-5 one by one (if there is a "1", the output is "1"; if all inputs are "1", the output is "0", etc.); analyze the difference between CMOS and TTL gate circuit in high and low output level; summarize the phenomenon of input suspension of CMOS gate circuit and TTL gate circuit.

4. Precautions

(1) Turn off the power supply before inserting the device into the socket or before removing the device from the socket, and turn it on only when test is carried out.

(2) The positive end and the ground end of the power supply cannot be wrongly connected.

(3) Inserted into or removed from sockets, devices should be handled horizontally, and must not be leveled out unilaterally, so as not to damage the pin of the devices, or use special tools to remove the devices.

5. Notes and summary

Task 2 Functional Test of Seven-Segment Code Locking/Decoding/Driver

【Relevant Knowledge】

Figure 4-10 shows the pin arrangement of the common cathode LED digital tube, Figure 4-11 shows the appearance and pin diagram of seven-segment latch decoder 4513, and the wiring of the digital tube and the 4513 device is shown in Figure 4-12.

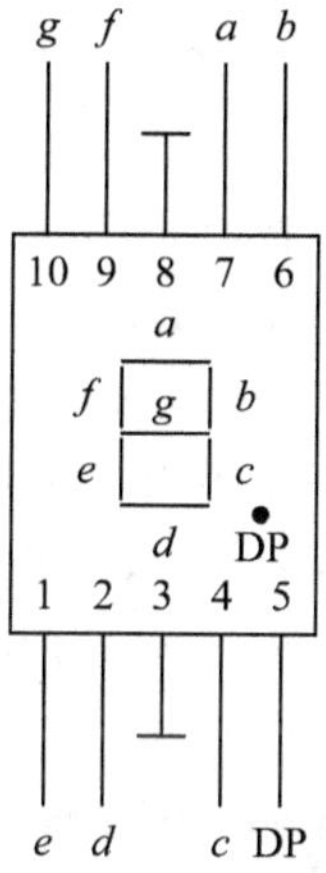

Fig. 4-10 Pin arrangement of digital tube

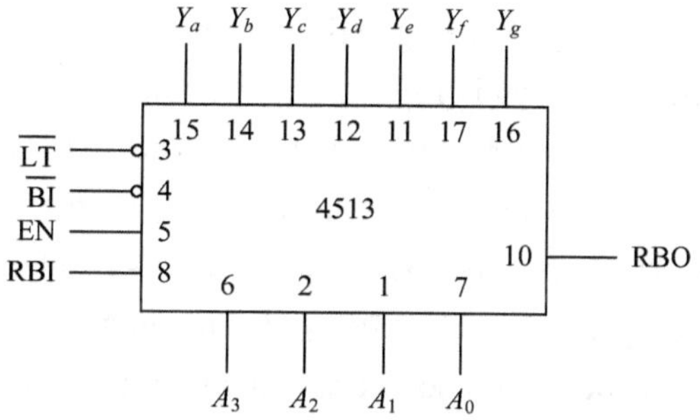

Fig. 4-11 Pin arrangement of 4513 device

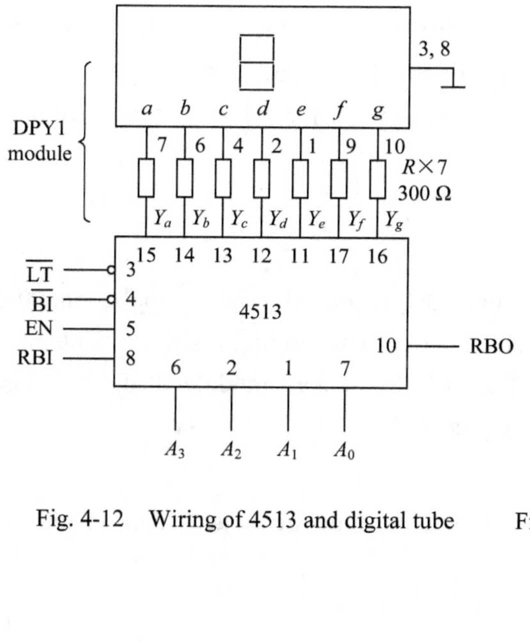

Fig. 4-12 Wiring of 4513 and digital tube

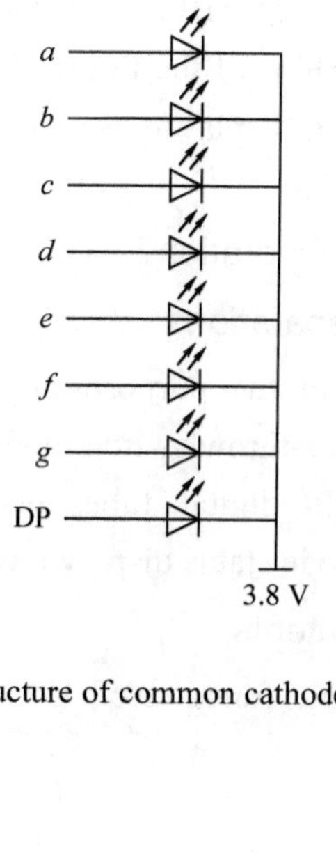

Fig. 4-13 Structure of common cathode LED digital tube

The cathode's NO.3 and NO.8 pins of eight light emitting diodes are connected and then grounded. Eight light-emitting diodes are arranged into seven digital segments and decimal points to lead out eight pins of a-g and DP. When anode of light emitting diode is connected to the high-level voltage, LED will be lighted. Therefore, when a-g pins are connected to the relative high-level voltage, digital tube will show numbers. The forward voltage drop of LED is $U_F \approx 2$ V and the forward current of the light-emitting diode is $I_F \approx 20$ mA. If the 5 V power is used to drive LED digital tube, a current limiting resistor of $R = \frac{5\,\mathrm{V} - U_F}{I_F} \approx \frac{(5-2)\,\mathrm{V}}{(20\times 10^{-3})\,\mathrm{A}} \approx 300\,\Omega$, should be connected in series, shown in Figure 4-12.

For COMS 4513 (as shown in Figure 4-11), its input A_3-A_0 is the binary-decimal BCD four-digit number 10(0-9), that is, 2(0000-1001). When a BCD code is input, the code segment of the corresponding number will be translated into high level for the digital tube to display the corresponding number. If the input A_3, A_2, A_1, and A_0 is 0101=5(decimal number), then the corresponding code segment *afgcd* is high level (if U_{DD} = +5 V, then *afgcd* are all 5 V), for the digital tube to display. However, 4513 also has 5 control function pins, with $\overline{\mathrm{LT}}$, $\overline{\mathrm{BI}}$, EN, and RBI as input and RBO as output. $\overline{\mathrm{LT}}$ is the test lamp input with top priority. Namely $\overline{\mathrm{LT}} = 0$, *a-g* are all 1, all LED are lighted in order to test whether the digital tube is intact. $\overline{\mathrm{BI}}$ is for off light input, with second priority. That is, when $\overline{\mathrm{LT}}$=1 and $\overline{\mathrm{BI}}$=0, *a-f* are all 0, no display by the digital tube. It should be noted that $\overline{\mathrm{LT}}$ and $\overline{\mathrm{BI}}$ are both set to 1 if other functions of the pins are to be used. EN being a latch, when EN=1, the digit before EN=1 is shown (that is, EN=0), and remains unchanged, and the new A_3-A_0 input is invalid. Only when EN=0, can the corresponding number be displayed with input. In addition, when the input of A_3-A_0 is from 10-15, the decoder does not work and without any display. Only when the input range of A_3-A_0 is 0-9, does the decoder work and the digital tube display relative numbers. The decoder also has a feature that when many 4513 chips are cascaded, the chips of the highest bit can be sent an input of A_3-A_0=0000. When RBI=1, the standard 0 cannot be displayed, and the output RBO=1 can be cascaded to the chips of the lower bit. If the chips of the low bit do not have an input of 0, the number 1-9 can be displayed. So when RBI=0, digits 0-9 can be displayed (4511 type have only RBI, RBO two pins).

【Practice and Operation】

1. Task preparations

According to the purpose of this training, one DC regulated power supply, one 0-1 single-pole double-throw switch module AX21, one DPY1 current-limiting resistor module, one common cathode digital tube, one double-in-line 18-pin socket module IC4, and one light-emitting diode status display module AX26 are prepared.

2. Task contents

Test the logic function of 4513 7-segment code latch/decode/driver according to the pin function status in Table 4-6. The output state of 4513 7-segment code latch/decode/driver is represented by the display number of common cathode light-emitting diode digital tube driven by current-limiting resistance *R*.

Table 4-6 4513 7-segment code latch/decoder/driver input/output function

No.	Input								Output								Functional
	$\overline{LT}$	$\overline{BI}$	EN	RBI	A_3	A_2	A_1	A_0	RBO	Display	Y_a	Y_b	Y_c	Y_d	Y_e	Y_f	analysis
1	0	0	×	×	×	×	×	×									
2	0	1	×	×	×	×	×	×									
3	1	0	×	×	×	×	×	×									
4	1	1	0	0	0	1	0	1									
5	1	1	0	0	1	0	0	1									
6	1	1	0	0	0	0	1	1									
7	1	1	0	0	1	1	0	0									
8	1	1	0	0	1	0	1	0									
9	1	1	0	0	0	1	1	1									
10	1	1	1	1	0	1	0	0									
11	1	1	0	0	0	1	0	0									
12	1	1	0	0	0	0	0	0									
13	1	1	0	0	0	0	0	1									
14	1	1	1	1	0	0	1	0									

3. Task procedures

(1) After switching on the power supply, the output voltage of the DC regulated power supply is adjusted to +5 V, and then the switch is closed.

(2) According to the circuit shown in Figure 4-12, insert the device 4513 and the LED digital tube into the corresponding socket and connect the current-limiting resistance. Connect 4513 input control terminal and digital input terminal with AX21 module 0-1 switch, and Y_a-Y_g terminal, RBO output, and AX27LED status monitor. Connect 4513, AX21 and AX26 to the power source respectively.

(3) When the regulated power supply switch is turned on, use the AX21 module to set the input state by serial number, and observe the digits displayed by the digital tube and the state of Y_a-Y_g. Make a record in Table 4-6.

(4) Turn off the regulated power supply switch and remove the wires and devices after the training.

4. Precautions

(1) There are many connections in this task. We should be careful when carrying out the connection so as to avoid connection errors.

(2) When connecting, connection should be done in the order of input and output in Table

4-6, which is beneficial to reading and recording.

5. Notes and summary

____

____

____

____

Task 3 Functional Test of Bidirectional Shift Register

【Relevant Knowledge】

The following is the working principle of bidirectional shift register.

Figure 4-14 is the logic symbol diagram of CD40194 (74LS194 has the same function) four-bit bidirectional shift register. There are 16 pins. The 16th pin and the 8th pin are connected to the positive end of the power supply and the ground respectively. The bidirectional shift register means that it can register the digits, which come from the four input terminals of D_0-D_3, and can register the corresponding digits in Q_0-Q_3. The bidirectional shift register also means that Q_0-Q_3 can move to the right or left bit by bit under the action of CP pulse. DSR can move to the Q_0 when moving to the right, while D_{SL} can move to Q_3 when moving to the left. The above functions are controlled by M_1M_0 state.

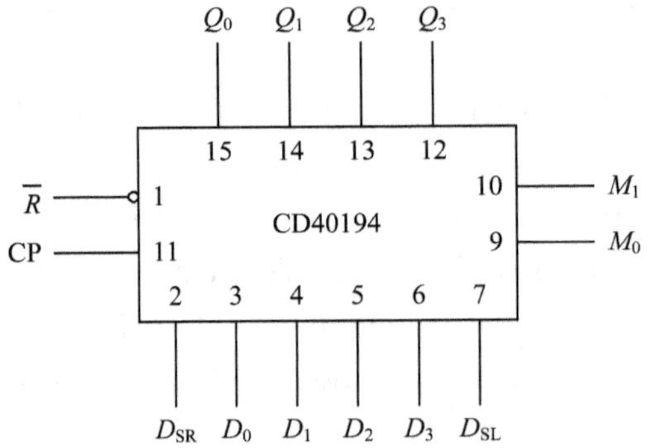

Fig. 4-14 CD40194 (74LS194 as well) logic symbol diagram of four-bit bidirectional shift register

【Practice and Operation】

1. Task preparations

(1) Preview the guidance materials of this training task and master the basic working principle.

(2) Prepare the equipment module for the training, that is, CD40194 bidirectional shift register.

2. Task contents

(1) Test the function of CD40194 bidirectional shift register as shown in Table 4-7.

Table 4-7 Function of CD40194 bidirectional displacement register

No.	Reset	Control signal		Clock	Parallel input		Parallel input				Output				Function
	$\overline{R}$	M_1	M_0	CP	D_{SR}	D_{SL}	D_0	D_1	D_2	D_3	Q_0	Q_1	Q_2	Q_3	
1	0	×	×	×	×	×	×	×	×	×	0	0	0	0	Reset
2	1	×	×	0	×	×	×	×	×	×	Q_0	Q_1	Q_2	Q_3	Keep
3	1	0	0	×	×	×	×	×	×	×	Q_0	Q_1	Q_2	Q_3	Non-displacement
4	1	0	1	↑	1	×	×	×	×	×	1	Q_0	Q_1	Q_2	Right shift
5	1	0	1	↑	0	×	×	×	×	×	0	Q_0	Q_1	Q_2	Right shift
6	1	1	0	↑	×	1	×	×	×	×	Q_1	Q_2	Q_3	1	Left shift
7	1	1	0	V	×	0	×	×	×	×	Q_1	Q_2	Q_3	0	Left shift
8	1	1	1	↑	×	×	d_0	d_1	d_2	d_3	d_0	d_1	d_2	d_3	Set number

(2) Construct pulse sequence generator and test its function.

The circuit of pulse sequence generator composed of two-way shift register circuit is as shown in Figure 4-15. Its working principle is as follows. When the starting signal of G_1 is 0, M_1=1, while M_0 is constant at 1, so M_1M_0=11 is set function. Under the action of CP pulse signal, $Q_0Q_1Q_2Q_3 = D_0D_1D_2D_3$=0111. If Q_0=0, G_2 comes out at 1, and the start signal is restored to 1, G_1 goes out 0, making M_1M_0=01, performing the right shift function. When a pulse signal is sent as an input, a shift occurs from Q_0=0 to Q_1=0, and Q_3=1 connects to D_{SR}, making Q_0=1. Because there is always a 1 in Q_0-Q_3, M_1 maintains 0. And when Q_3=0 connects to D_{SR}=0, then another pulse signal comes, and Q_0 is 0. Therefore, under the action of the pulse signal, the right shift function is continuously carried out, Q_0-Q_3 take turns to output low level, and its output waveform is shown in Figure 4-16.

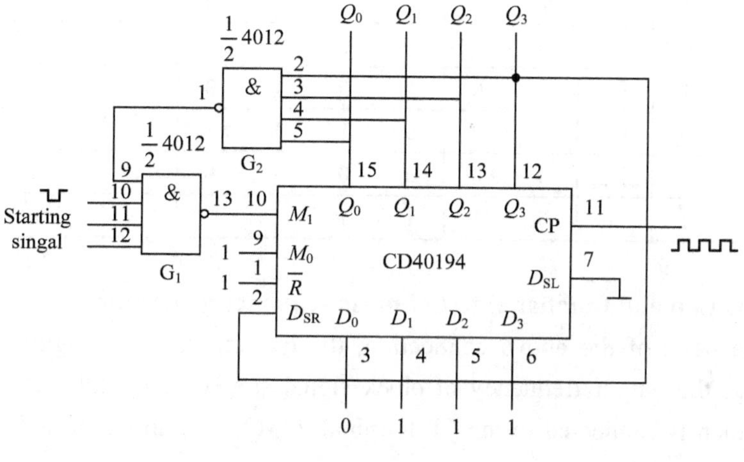

Fig. 4-15 Circuit diagram of pulse sequence generator

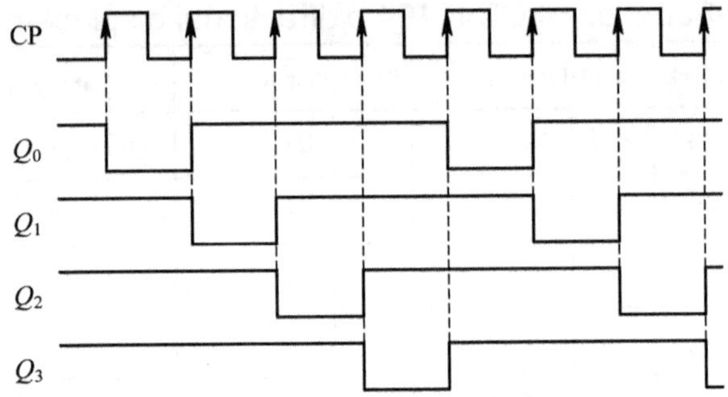

Fig. 4-16　Input and output waveform of pulse sequence generator

3. Task procedures

1) Functional test of CD40194 bidirectional shift register

(1) Turn on the DC regulated power supply and adjust the output voltage to + 5 V, and then turn off the power supply.

(2) Insert the CD40194 device into the 16-pin socket of IC3.

(3) Connect input control, and input digital terminal M_1, M_0, R, D_0-D_3, D_{SR}, and D_{SL} (9 in total) with 0-1 state settings of AX21, and connect the CP terminal with AX22 single-phase pulse signal source.

(4) Connect Q_0-Q_3 output with AX26 LED status display lamp.

(5) Connect the power supply of each device and module with the stabilized power supply.

(6) Turn on the power supply according to Table 4-8, the sequence number shown is tested in turns, and fill in the following table.

Table 4-8　Functional test of bidirectional shift register

No.	Input										Output				Logic function
	$\overline{R}$	M_1	M_0	CP	D_{SR}	D_{SL}	D_0	D_1	D_2	D_3	Q_0	Q_1	Q_2	Q_3	
1	0	1	0	1	0	1	1	1	1	1					
2	1	1	1	⎍	1	0	1	0	1	0					
3	1	1	0	⎍	0	1	1	0	1	0					
4	1	0	1	⎍	1	0	1	0	1	0					
5	1	0	0	⎍	1	0	0	1	1	1					
6	1	1	1	⎍	0	0	0	0	1	0					
7	1	0	1	0	0	1	1	1	0	1					

2) Construction and functional test of pulse sequence generator

(1) On the basis of the above connection, the circuit shown in Figure 4-15 is rewired. Before rewiring, the output frequency of clock signal source of module AX25 is adjusted to 0.5-1 Hz, and then is connected to the CP terminal. Q_0-Q_3 are still connected with AX26 LED status indicator.

(2) The “start signal” terminal of G_1 NAND gate is connected with the “output” terminal of

AX22 single-pulse signal generator "⅂⅃⌐ "and connected with power supply.

(3) Turn on the regulated power supply, press the button of AX22 single-pulse signal source, and observe the change of the CP end and the LED status display lamp indicating the states of Q_0-Q_3.

(4) Record the state of CP and observe the state of Q_0-Q_3 at the same time.

(5) Turn off the regulated power supply after the training.

4. Precautions

(1) There are two kinds of signals of AX22 single-pulse signal generator connected to CP clock terminal in the "task contents(1)". The signals of AX22 single-pulse signal generator are "⅂⅃⌐" and "⌐⅂⅃", the former indicating that the static state is 0, and the pressed button is 0 →1→ 0 of single-phase positive pulse, the latter indicating that the static state is 1, and the pressed button is 1→ 0 →1 of single-phase negative pulse. So the single-phase negative pulse should be selected according to the need.

(2) The CP clock signal generator of the "task contents (2)" requires that the frequency f should be about 0.5-1 Hz, which can be adjusted in low frequency band. It is not necessarily required to be exactly correct, and can be judged by vision.

5. Notes and summary

____

____

____

____

【Extension Exercises】

Exercise 1 Functional Test of Binary-Quinary-Decimal Counter

1. Task objectives

(1) Master the basic working principle of 74LS196 counter chip.

(2) Build a binary-quinary-decimal counter test circuit and test its technical functions.

2. Equipment

DC stabilized power supply, 0-1 status setting switch module AX21, LED status display module AX26, single-phase pulse signal generator module AX22, and 14-pin socket module IC2. Training device: 74 (54) LS196 two-five decimal counter.

3. Task contents

Figure 4-17 is the logic symbol diagram of 74LS196 binary-quinary-decimal counter. The so-called counter counting is used to count the number of CP pulses, while the basic unit circuit of the counter is a trigger of various clock triggering operations. The state of each output Q of multiple triggers represents a binary number, which can form counters of various counting systems. The device shown in Figure 4-17 is divided into two independent circuits. Q_0 indicates that one-bit binary counting output is triggered by the falling edge of $\overline{CP_0}$ pulse for Q_0 to change. Q_3-Q_1 indicates that the three-bit binary state is effective for $\overline{CP_1}$ pulse counting and is triggered by the falling edge. However, the circuit is composed of a quinary counting system, that is, the counting sequence from 000 to 100 goes through five counting state cycles. The state reaches 100 and then returns to 000. Table 4-9 is the function table of the device.

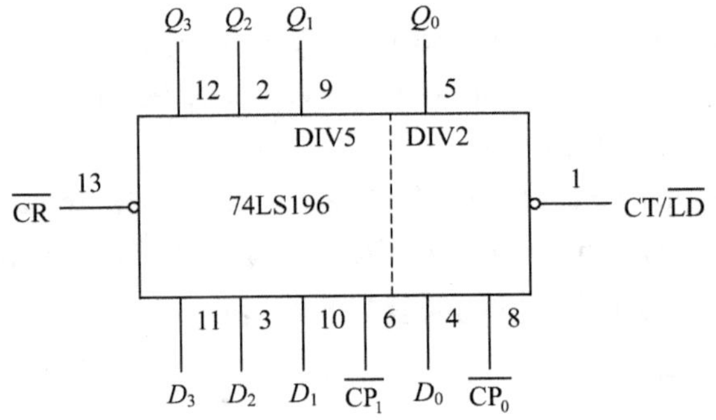

Fig. 4-17　Logic symbol diagram of 74LS196 binary-quinary-decimal counter

Table 4-9　74LS196 functional table of binary-quinary-decimal counter

Input								Output				Logic function
$\overline{CR}$	CT/$\overline{LD}$	$\overline{CP_0}$	$\overline{CP_1}$	D_3	D_2	D_1	D_0	Q_3	Q_2	Q_1	Q_0	
0	×	×	×	×	×	×	×	0	0	0	0	Asynchronous zero clearing
1	0	×	×	d_3	d_2	d_1	d_0	d_3	d_2	d_1	d_0	Set number
1	1	↓	×	×	×	×	×	Keep			Count	Binary addition counting
1	1	×	↓	×	×	v	×	Count			Keep	Quinary addition counting

4. Task procedures

The 74LS196 device is inserted into the IC2 socket. Six input terminals which are D_3-D_0, $\overline{CR}$, and CT/$\overline{LD}$, are connected with AX21 module respectively. Four output terminals of Q_3-Q_0 are connected with the status display of the module AX26, while $\overline{CP_0}$ and $\overline{CP_1}$ are connected with AX22 module of single-phase pulse generator respectively. Adjust the DC regulated power supply to +5 V. Connect each module and device after switching off the power supply, and input the pin states one by one by the serial numbers listed in Table 4-10. Observe the Q_3-Q_0 states and record them in the table.

Table 4-10　Functional test of 74LS1962 binary-quinary-decimal counter

No.	Input								Output				Logic function
	$\overline{CR}$	$CT/\overline{LD}$	$\overline{CP_0}$	$\overline{CP_1}$	D_3	D_2	D_1	D_0	Q_3	Q_2	Q_1	Q_0	
1	0	1	1	0	1	0	1	1					
2	1	0	1	⎍	1	0	1	1					
3	1	1	1	⎍	1	0	1	1					
4	1	1	1	⎍	1	0	1	1					
5	1	1	1	⎍	1	0	1	1					
6	1	1	1	⎍	1	0	1	1					
7	1	1	1	⎍	1	0	1	1					
8	1	1	1	⎍	1	0	1	1					
9	1	1	⎍	1	1	0	1	1					
10	1	1	⎍	1	1	0	1	1					
11	1	1	⎍	1	1	0	1	1					
12	1	1	⎍	1	1	0	1	1					

5. Precautions

(1) Turn off the power supply before the device is inserted into the socket or taken out of the socket, and turn on the power supply when testing.

(2) The positive terminal of the power supply and the grounding terminal should not be wrongly connected.

(3) The device should be inserted or taken out horizontally when it is inserted into or taken out of the socket. It is not allowed to pry it out unilaterally to avoid damaging the pin of the device. Or take out the device with special tools.

Exercise 2　Functional Test of Priority Encoder

1. Task objectives

(1) Test its logic function through the use of the functional table of CD4532 priority encoder, and learn to understand the pin function in the function table so as to use integrated devices.

(2) Understand the meaning of priority coding of priority encoder.

(3) Understand the control function of EI, GS and EO pins of CD4532 priority encoder.

(4) Understand the numerical relationship between CD4532 priority encoder input I_i and coded outputs Y_2, Y_1, and Y_0.

2. Equipment

Regulated power supply, two-in-line insertion module IC3, 0-1 set single-pole double-throw switch module AX21, and light-emitting diode drive circuit module AX26. Training device: CD4532.

3. Task content

(1) Master the basic working principle of CD4532 priority encoder.

CD4532 priority encoder is a 16-pin integrated chip, and Figure4-18 is its logic symbol diagram. The U_{DD} of pin 16 of the device is connected to the power supply of +5 V, and pin 8 is grounded.

CD4532 is a combined logic circuit, which is composed of gate circuits. Its output logic state changes with the input logic state. CD4532 priority encoder requires that the input coding pin should be I_7-I_0, the highest coding priority being I_7, the lowest being I_0, and the corresponding coding output being Y_2, Y_1, and Y_0, which is the meaning of priority coding, that is, using Y_2, Y_1, and Y_0 three-bit binary code to represent the specific information of priority I_i.

Generally, the integrated combined logic device is equipped with enabling terminal pins to control whether the device can work, that is, the function of EI. For the chip, its priority level is the highest (not coding priority). EI being 0 means the prohibition of work while EI being 1 means the permission of work.

(2) According to the status of each input pin listed in Table 4-11, test the logic status of each output pin.

Table 4-11 Training test content

No.	Input									Output				
	EI	I_7	I_6	I_5	I_4	I_3	I_2	I_1	I_0	Y_2	Y_1	Y_0	GS	EO
1	0	0	0	1	1	1	1	0	1					
2	1	0	0	0	0	0	0	0	0					
3	1	0	0	1	1	1	0	0	1					

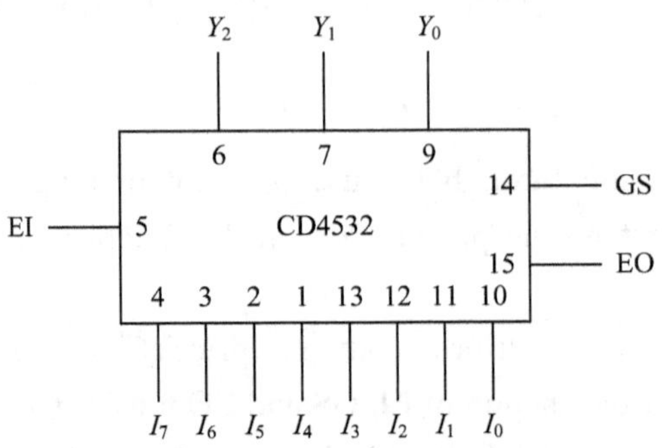

Fig. 4-18 Logic symbol diagram of CD4532 priority encoder

4. Task procedures

(1) Turn off the power switch after adjusting the output DC voltage of the regulated power supply to +5 V.

(2) Insert the CD4532 device horizontally into the 16-pin socket of IC3 to connect the power supply.

(3) The 0-1 switch of AX21 module is connected to the pins of EI and I_0-I_7 of CD4532. And connect the power source of +5 V.

(4) Five output terminals of CD4532 are connected to the input hole of the light-emitting diode of AX26 and connected to the +5 V power supply line of AX26.

(5) Set the 0 and 1 states of AX21 according to the serial number 1 in"Training Test Content Table".

(6) Turn on the voltage regulator, observe the five output pins of CD4532 and record them in Table 4-11. If the LED light-emitting diode in AX26 is "on", it means 1 state, and if the LED light-emitting diode in AX26 is "off", it means 0 state.

(7) According to the input state of the serial number 2 in Table 4-11, the setting state of each AX21 0-1 switch is changed accordingly, and then the output pin state is observed and recorded in the table.

(8) The contents of serial number 3 in the content table are tested by Step(7).

(9) Hold any input pin vacant or poor contact, and observe the output state.

(10) After the test, turn off the voltage regulator switch first, then remove all the connections, and take out the CD4532 device horizontally.

5. Precautions

(1) The pin arrangement diagram of CD4532 can be referred to the pin arrangement of devices in the appendix. In the digital integrator, the pin number is not marked on the logical symbol diagram, so the pin arrangement diagram must be consulted.

(2) Reliable contact is required for each connection.

(3) Do not pull out or rewire the pin of the device when the power supply is switched on.

Exercise 3 Functional Test of Binary Decoders and Data Selectors

1. Task objectives

(1) Be familiar with the meaning and logic function of binary decoder and data selector.

(2) Understand the logical function tables of 74HC138 3-8 line decoder and 74HC151 1in 8 data selector (Readers should consult books for the relevant information by themselves), and

master the pin function.

(3) Grasp the correct usage method of 74HC138 and 74HC151 and the logic regularity of input and output.

2. Equipment

DC regulated power supply, double in-line 16-pin module IC3; 0-1 set single-pole double-throw switch module AX21, light-emitting diode and drive circuit module AX26, and training devices 74HC138 and 74HC151.

3. Task contents

(1) Master the basic working principle of decoder and data selection.

It can be seen from Figure 4-19 that the input is 3-bit binary codes A_2, A_1 and A_0, and there are 8 combinations from 000 to 111. When a group of code is input (also known as address code), it is possible to translate a certain effective low-level signal $\overline{Y}_i$ output accordingly. There are 8 corresponding output signals for 8 address codes, and there is a corresponding relationship between the decimal numbers of the address code, which is called 3-8 line binary decoder. In addition, there are three enabling control terminals E_3, $\overline{E}_2$ and $\overline{E}_1$. From the logical symbols, only if E_3=1, $\overline{E}_2$=0, and $\overline{E}_1$=0, can the decoder work, or otherwise the coding cannot be carried out. Binary encoder, also known as address code decoder, is the basic circuit in semiconductor memory.

74HC151 is a 1 in 8 data selector. Seen from Figure 4-20, it has eight data input terminals D_0-D_7. The data output is determined by the address code of three digits A_2, A_1 and A_0. The subscript i of D_i represents the corresponding relationship between the decimal number and the address code decimal system. The output Y is the original code of D_i, the output $\overline{W}$ is the inverse code of D_i, and the $\overline{\text{ST}}$ is the chip selection control terminal. If $\overline{\text{ST}}$ is low level, work is done. If $\overline{\text{ST}}$=1, work is prohibited.

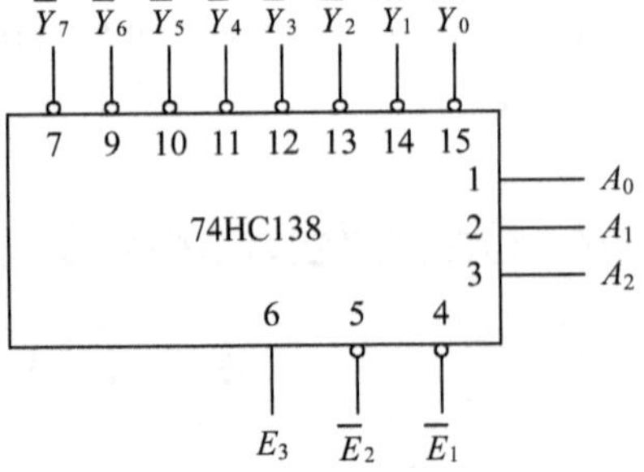

Fig. 4-19　3-8 Logical symbol diagram of line decoder

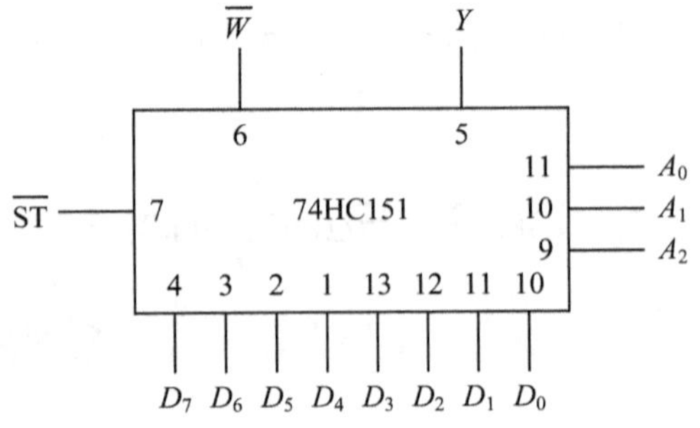

Fig. 4-20　Logical symbol diagram of 1in 8 data selector

(2) Test the logical function of 74HC138 3-8 line decoder. Fill testing results into Table 4-12. And then test the logical function of 74HC151 1in 8 data selector.

Table 4-12 Logical function test of 74HC138 3-8 line decoder

Enabling input			Address code decoding input			Decoding output							
E_3	$\overline{E}_2$	$\overline{E}_1$	A_2	A_1	A_0	$\overline{Y}_0$	$\overline{Y}_1$	$\overline{Y}_2$	$\overline{Y}_3$	$\overline{Y}_4$	$\overline{Y}_5$	$\overline{Y}_6$	$\overline{Y}_7$
0	×	×	×	×	×								
×	1	×	×	×	×								
×	×	1	×	×	×								
1	0		0	0	0								
1	0		0	0	1								
1	0		0	1	0								
1	0		0	1	1								
1	0		1	0	0								
1	0		1	0	1								
1	0		1	1	0								
1	0		1	1	1								

4. Task procedures

(1) Turn the DC regulated power supply on and adjust it to +5 V and turn off the power supply. The power supply front end and ground end are connected to the power supply end of training equipment AX21, AX26 and IC3 respectively.

(2) The integrated circuit chip 74HC138 is inserted horizontally into the IC3 socket.

(3) Connect the 0-1 switch of the AX21 module to E_3, $\overline{E}_2$, $\overline{E}_1$ and A_2, A_1, A_0 input terminals respectively, and connect its 8 output terminals Y_0-Y_7 to AX26 socket holes respectively.

(4) Test the decoded output state according to the input state listed in Table 4-12, test after switching on the regulated power supply, and fill in Table 4-12. After setting the input state with 0-1 switch, observe the output LED, on indicating 1, and off indicating 0.

(5) After switching off the power supply, the chip is replaced by 74HC151. Because AX21 has only 10 sockets, the unused D_i can be connected in parallel.

(6) When the power supply is turned on, observe and record the output state of Y and $\overline{W}$.

(7) After the training, turn off the power supply and disconnect the wiring and devices.

5. Precautions

(1) Do not install or remove devices or wiring when power is on.

(2) Device installation and removal should be in a horizontal state. It cannot be leveraged unilaterally to avoid pin damage.

(3) When testing, the power supply should be turned on after all the connections have been completed.

(4) If the logic state of output is not normal, first of all, the DC voltage of the multimeter

can be used to detect whether the power supply voltage is normal, and then touch the surface of the device to check whether the surface of the device is of high temperature. If there are no problems above, then check whether the logic level voltage set by each input pin is correct and whether the contact is good. The DC voltage of a multimeter can be used to detect its level voltage, and then check the correctness of the DC level at the output end one by one to determine whether the device is damaged.

Exercise 4　Functional Test of Full Adder and Carry-Ahead Full Adder

1. Task objectives

(1) Understand the logical functions of full adder and carry-ahead full adder.

(2) Master the usage of 74LS283 carry-ahead full adder.

(3) Test the logical functions of full adder composed of integrated logic gates and 74LS283 carry-ahead full adder.

2. Equipment

DC regulated power supply, 0-1 set-up single-pole double-throw switching module AX21, light-emitting diode and driving circuit module AX26, double-in-line socket module IC3 and IC2, the training devices CD4070, CD4085, and 74LS283.

3. Task contents

(1) Master the basic working principle of full adder.

Full adder is an arithmetic operation logic circuit for binary number addition. It uses logical states 1 and 0 to represent one binary number. If two binary numbers A and B are used for addition operation, the circuit that does not participate in addition operation for low bit carriers is called semi-adder. And if low bit carriers also participate in addition operation, then this circuit is called full adder. Figure 4-21 is a unit circuit of full-adder consisting of logic gates. The inputs are A_i and B_i are for augend and addend, C_{i-1} is for carry at low bit, and the outputs S_i and C_i are for bit and carry at high bit respectively. Their logical states 1 and 0 are binary. $S_i = A_i \oplus B_i \oplus C_{i-1}$ where A_i, B_i and C_{i-1} are an odd number of 1, and the sum S_i must be 1. In the figure, G_3 XOR gate is used as an inverter. That is, when one of its input terminals is set to 1, the output must be opposite to the other input state by the rule of the output 0 if the same and the output 1 if different. Therefore, $C_i = A_i \cdot B_i + (A_i \oplus B_i)C_{i-1}$ meaning that A_i and B_i are both 1 or when C_{i-1} is 1, if one of A_i and B_i is 1, the carry C_i must be 1. This is the logical state relationship between the output and the input value of the full adder. 74LS283 (or CD4008) of 4-bit binary carry-ahead full adder is able to realize the addition of multi-bit binary calculation. Its logic symbol diagram of 74LS283 is shown in Figure 4-22.

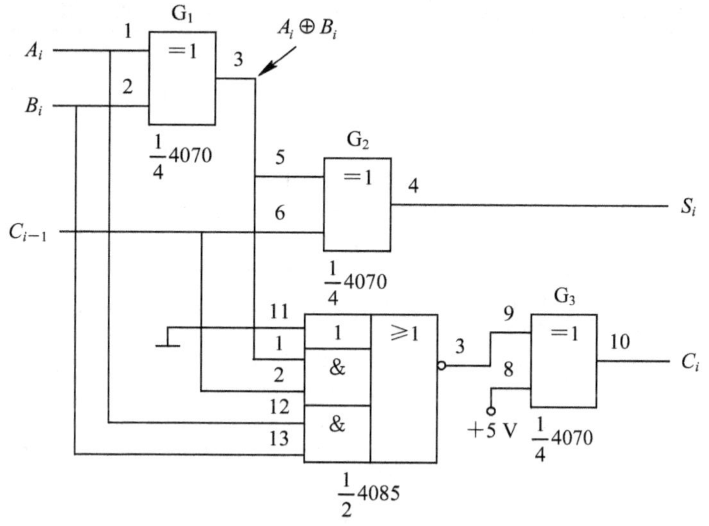

Fig. 4-21 Gate circuit of full adder

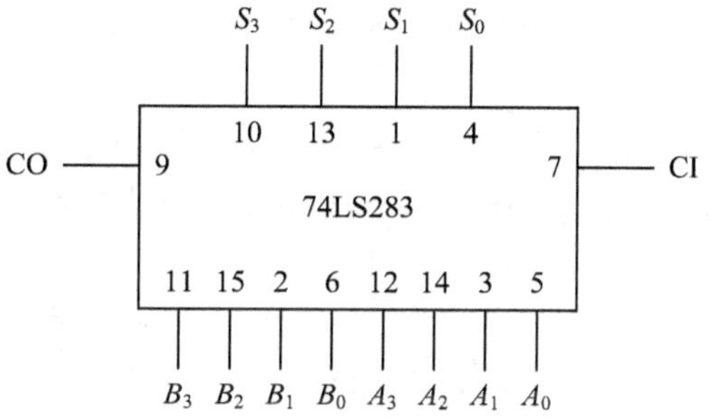

Fig. 4-22 Logic symbol diagram of 74LS283 carry-ahead full adder

(2) Test the output logic status of the gate circuit of full adder and 74LS283 4-bit binary carry-ahead full adder, and fill the test results in Tables 4-13 and Table 4-14.

Table 4-13 Test of gate circuit of full adder

Input			Output	
Augend A_i	Addend B_i	Carry at low bit C_{i-1}	Standard sum S_i	Carry at high bit C_i
0	0	0		
0	0	1		
0	1	0		
0	1	1		
1	0	0		
1	0	1		
1	1	0		
1	1	1		

Table 4-14 Test of 74LS283 carry-ahead full adder

Input											Output				
$(N_A)_{10}$	A_3	A_2	A_1	A_0	$(N_B)_{10}$	B_3	B_2	B_1	B_0	CI	CO	S_3	S_2	S_1	S_0
$(2)_{10}$	0	0	1	0	$(3)_{10}$	0	0	1	1	1					
$(6)_{10}$	0	1	1	0	$(9)_{10}$	1	0	0	0	0					
$(14)_{10}$	1	1	1	0	$(11)_{10}$	1	0	1	1	1					
$(10)_{10}$	1	0	1	0	$(8)_{10}$	1	0	0	0	0					
$(15)_{10}$	1	1	1	1	$(12)_{10}$	1	1	0	0	1					

4. Task procedures

(1) Turn on the DC regulated power supply and adjust it to +5 V, then turn off the power supply.

(2) Insert the devices CD4070 and CD4085 into the IC2 socket horizontally, and connect the power supply.

(3) Connect inputs A_i, B_i, and C_{i-1} with AX21 module to set 0 and 1 states, and connect outputs S_i and C_i with AX26 module to observe output 0 and 1 states, and connect AX21 and AX26 modules with power supply.

(4) Turn on the voltage regulator switch, and then observe the output states displayed by AX26. The LED light on indicates 1, and the LED light off indicates 0, and record LED states in Table 4-13.

(5) After the test, turn off the voltage regulator switch and remove the device and connection.

(6) Insert 74LS283 device into IC3 socket horizontally to connect power supply.

(7) Connect input A_3-A_0, B_3-B_0 and CI pins with AX21 module 0-1 switch, and connect output CO, and S_3-$S_0$with AX26 module and power supply.

(8) Turn on the regulated power supply, set the input state of each line with AX21 module, and then observe the output state shown by AX26 module. The LED light on is 1 and off state is 0. Fill the output status in Table 4-14.

(9) After the completion of the training and testing, turn off the voltage regulator switch and disconnect the devices and connections.

5. Precautions

(1) Do not install or disconnect devices or connections when power is turned on in order to avoid damaging to integrated devices.

(2) When installing or unplugging devices, it is necessary to maintain a horizontal state and avoid unilateral warping and pin damage.

Exercise 5 Functional Test of Numerical Comparator

1. Task objectives

(1) Be familiar with the logic function of numerical comparator.

(2) Be familiar with 4-bit digital comparator CD4585 pin function and usages.

(3) Master the connection method of 2-bit decimal number made of 4-bit numerical comparator and dial.

2. Equipment

DC regulated power supply, AX24 dial modules, IC3 16-pin sockets, AX26 LED, state display module and training devices CD4585.

3. Task contents

(1) Master the basic logic function of numerical comparator.

Figure 4-23 is a 1-bit binary numerical comparator consisting of gate circuits. Since the binary numbers A and B are only 1 and 0, there are 3 output states. ① When A=1 and B=0, that is, $A>B$, the output Z_2 ($A>B$) is 1 state, and its output is 0 state. ② When A=0 and B=1, that is, $A<B$, the output Z_1 ($A<B$) is 1 state, and the other outputs are 0 state. ③ When $A=B$, the output Z_3 ($A=B$) is 1 state, the other outputs are 0 state.

The 4-bit binary numerical comparison circuit is shown in Figure 4-24 which consists of two 4585 4-bit numerical comparators. The 4-bit binary data are from 0000 to 1111. Now the circuit is used to compare 2-bit BCD code decimal number. A_3-A_0 and B_3-B_0 are the input terminals of two 4-bit binary numbers A and B. The comparison results are output by three terminals $A<B$, $A=B$ and $A>B$. If the highest bit of A and B is $A_3>B_3$, the output ($A>B$) is 1, while the outputs ($A=B$), ($A<B$) are 0. If the highest bit $A_3<B_3$, the output ($A<B$) is 1 and the rest outputs are 0. If the highest bit $A_3=B_3$, the output result is determined by judging the sub-high bit A_2 and B_2, and so on.

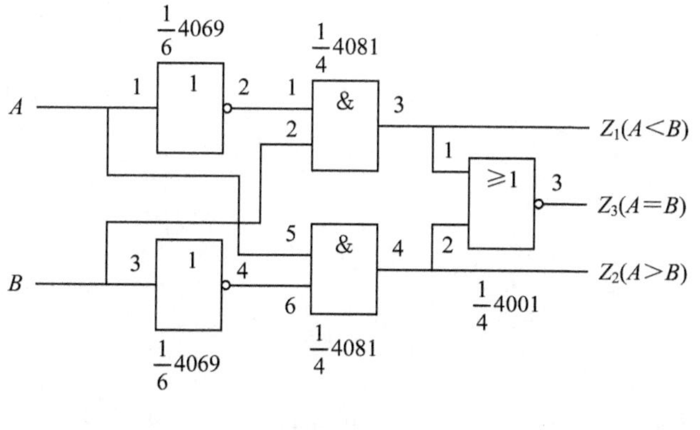

Fig. 4-23 1-bit binary numerical comparator

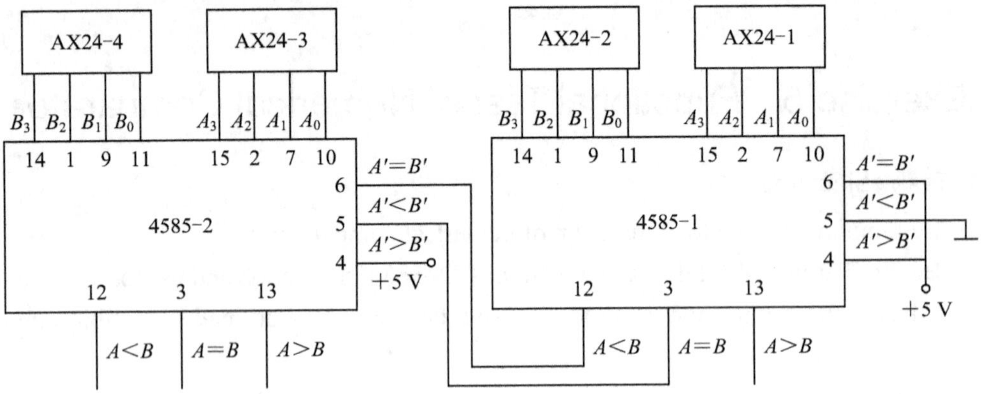

Fig. 4-24 2-bit BCD code decimal comparison circuit

4. Task procedures

(1) Decimal numbers of A and B are converted into 8421 BCD codes.

(2) Connect the device and module as shown in Figure 4-24, and connect the comparison outputs of 4585-1 and 4585-2 with AX26 LED display module. The dialing template of AX24 should be connected to the end marked with (set) font.

(3) Adjust the DC regulated power supply to +5 V, turn off the power supply, and connect the power supply to the power terminal of each device and AX24 module.

(4) Press the increasing or decreasing button of the dial to observe the displayed number of the dial. Turn on the power supply after setting the values with the dial for each serial number, and the output state of 4585 high and low bits are observed. A connection line of AX26 LED state display module is used to test the state of B_3-B_0 and A_3-A_0 terminals, and check whether they match the number of BCD codes.

5. Precautions

(1) The number disc on the AX24 8421 dial is decimal, and its output is 8421 BCD code.

(2) This training uses the setting function of AX24 8421 dial module.

Appendix

Appendix 1 Sample of Training Report

Training Report

Course Name:			Class:
Training Project:			Student Name:
Teacher		Score:	Date:
Training objectives			
Training equipment			
Training content			
Training procedures			
Test results			
Conclusion & summary			

Appendix 2　Standards for Safe Operation

1. Basic requirements for experimental training

1) Preparation before training

(1) Read the experimental guidebook carefully, review the relevant contents of the textbook and take notes. Preview knowledge well in advance, and clarify relevant theoretical knowledge. Otherwise, you are likely to have a half-baked knowledge, encounter difficulties and be at a loss.

(2) Make clear the purpose of the experiment, list out the items and steps of the experiment, and go to the laboratory in advance to understand the structure characteristics of the experimental device, the unit of the experiment and the necessary accessories and instruments.

(3) Make clear and remember the precautions of the experiment, avoid personal safety accidents and equipment damage. We must bear in mind that we should not be careless.

To sum up, the preparation before the experiment is the precondition for cultivating good habits and completing the experiment task.

2) Task procedures

(1) Establish a team and divide the work. Experiments are usually carried out in groups of two to three people. The experiments usually involve wiring, load adjustment, meter reading, data recording, etc.. The members of the experimental group should have a clear and reasonable division of labor, take their places, coordinate their work, and constantly consult in the experiment to ensure the standardization and coordination of the experimental operation.

(2) Select units and instruments. Before the connection in the experiment, the appropriate units and instruments (including the range and grade of the instruments) should be selected according to the purpose of the experiment, and their models and parameters should be recorded.

(3) Graphical wiring. The method for connection follows the rule of series first and parallel second. From the beginning to the end, connect components in series to make a main circuit, and then connect components in parallel one by one. It is better to use the wires of the same color for each route for convenient check. After the wiring is completed, it must be checked by another member before the experiment can be carried out.

(4) Reading and Recording. When reading instruments, attention should be paid to the ratio, minimum scale value, unit and effective number. If any abnormalities are found, the power supply should be cut off immediately, the causes should be found and the troubles should be eliminated.

(5) Removing wires. After the experiment, it is often the time when accidents most likely happen due to a relaxed mood. Therefore, the power must be cut off before removing the wires, and then wait for a short time to release the charge stored in the capacitor. And then remove the wires and put away the wires, instruments and experimental units.

3) Training Report

The experimental report is based on the measured data, the observed phenomena and the problems found in the experiment, and is analyzed, studied and discussed, so as to draw scientific conclusions. It trains students to process data and analyze graphics. Summarizing conclusions and writing scientific and technological documents are very important practices. Therefore, attention should be paid to the requirements for writing experimental reports. The experimental report requires clear thinking, concise writing, neat charts, reasonable analysis, clear conclusions, standardized writing and smoothness.

2. Operation rules for training safety

In order to ensure personal safety and equipment safety during the experiment, the experimental members must strictly observe the following safety operation rules.

(1) The experimenter should make sure that the experimental device has the reliable protection of grounding, leakage protection, overload protection, insulated desktop, carpet and safe wiring.

(2) During the experiment, the human body should not touch the bare charged body.

(3) When wiring and disconnecting, the power supply must be cut off.

(4) When switching on, all members of the experimental group must be clear of starting the experiment and agree to start it.

(5) In case of abnormal circumstances, such as smelling the odor (such as the burning smell of varnish), finding that the unit is overheated or vibrating too much (or running noise too much), that the fuse is burned out, and that sparks appear in the unit, and seeing that the meter exceeds the full range (or reverse deviation), the power supply should be cut off immediately to find out the causes and troubleshoot.

(6) When DC motor starts directly, it must be ensured that the excitation power supply is connected. If the excitation power supply is not connected directly to the armature power supply, it is easy to cause a flying accident. Secondly, the variable resistor should be connected in series in the armature circuit, and the sliding point of the variable resistor should be placed at the maximum resistance value. Otherwise, the armature circuit current will be too large (approximate short circuit), burning the motor.

(7) Before the unit is electrified, it is necessary to check whether the motor can rotate flexibly. Check whether the rotor is stuck. If the rotor cannot rotate flexibly, it should be readjusted so that the rotor can rotate flexibly.

(8) The total power supply of the laboratory should be controlled by the experimental instructor. Other personnel can only operate with the permission of the instructor. They are not allowed to switch on without permission. If they do so, serious accidents may occur.

(9) When using a resistor, it should be noted that the resistor should not be placed at zero (or very small resistance value), which is likely to burn out the rheostat. In the experiment, first adjust the high value resistance, and then adjust the low value resistance.

Appendix 3 Training Standards for Electrotechnology Skills (Elementary Level)

A Series of Textbooks for Lancang-MekongVocational Education Training Center and Cambodia Luban Workshop

Electrotechnology—Elementary Level
The Training Standard

Training hours: 96 hours
Personnel in charge: Zhang Jia, Zhang Lian, Zhang Xinjiang and Li Yunlong

Tianjin Sino-German University of Applied Sciences
May 2021

I. Overview of Training

According to the *National Vocational Qualification Standards for Electronics*, and in light of the employment needs of local people in Cambodia and Chinese-funded enterprises, technical skills training will be carried out by relying on the Lancang-Mekong vocational education and training center and experimental training equipment of Luban Workshop in Cambodia.

II. Objectives of Training

Local people in Lancang-Mekong countries and employees of Chinese-funded enterprises are engaged in or ready to engage in the installation, debugging, operation, maintenance and repair of mechanical equipment and electrical system lines and devices.

III. Design of Training

1. Overall of design

The elementary electrician training is designed to cover 96 hours, which is divided into four training modules: integrated training of the installation of basic electrical circuits, integrated training of electrical control circuit, integrated training of analog electronic technology and integrated training of digital electronic technology. Under each training module, two to four tasks are designed. Each task is planned to train for 8 hours, which covers the explanation of theoretical knowledge and practical operation training required by the task. At the same time, each training module provides extension training tasks for well-grounded students to change tasks or further practice after class.

2. Basis of design

According to the *National Vocational Qualification Standards for Electronics* and the local training needs and specific training conditions of electrical technology skills in Cambodia, students can master the basic theoretical knowledge and practical skills of electricians, and can use the knowledge and skills learned to solve the general problems related to electrical technology in production posts, and can independently analyze and solve the simple problems and faults in the experiment and acquire a knowledge of safe power use and civilized production.

3. Thoughts on content arrangement

This elementary training course adopts the teaching method of combining multimedia teaching with experimental teaching. According to the characteristics of each module, heuristic teaching runs through the whole course. Teachers are required to pay full attention to classroom teaching in the teaching process, focusing on how to guide students to analyze and solve problems in the face of an electrical system, combine imparting knowledge with developing ability, for students to deepen a real understanding of knowledge through independent learning, for them to improve professional technology ability by solving practical problems, professional working ability and professional accomplishment.

4. Class schedules

Training Module	Task	Credit Hours	Total
Module One: Comprehensive Training in Basic Electrical Circuit Installation	Task 1 Installation, Connection and Operation of Indoor Power Distribution and Lighting Control Circuit	8	16
	Task 2 Installation, Connection and Operation of the Standby Power Supply Circuit for Automatic Switching of Overcurrent in Power Grid	8	
Module Two: Comprehensive Training in Electrical Control Circuit	Task 1 Installation, Connection and Operation of Direct Start Control Circuit of Three-Phase Asynchronous Motor	8	32
	Task 2 Installation, Connection and Operation of Forward and Reverse Control Circuits of Three-Phase Asynchronous Motor	8	
	Task 3 Installation, Connection and Operation of Y-△ Start Control Circuit of Three-Phase Asynchronous Motor	8	
	Task 4 Installation, Connection and Operation of Direct Switching Control Circuit of Two-Speed Three-Phase Asynchronous Motor	8	
Module Three: Comprehensive Training in Analog Electronic Technology	Task 1 Performance Test of Semiconductor Diodes and Triodes	8	24
	Task 2 Basic Training of PCB Welding	8	
	Task 3 Manufacturing of DC Voltage Stabilizing Power Supply	8	
Module Four: Comprehensive Training in Digital Electronic Technology	Task 1 Functional Test of Basic Logic Gate Circuit	8	24
	Task 2 Functional Test of Seven-Segment Code Locking/Decoding/Driver	8	
	Task 3 Functional Test of Bidirectional Shift Register	8	
Credit Hours in Total	96		

Ⅳ. Training Objectives

(Ⅰ) Overall objectives

This project aims to design four typical training modules and 12 practical tasks for practitioners who are involved in electrician profession for the first time. The main training objectives are cognition, experience and basic operation. The contents include basic knowledge and skills of power electronics, basics of analog and digital electronics, power laying and electrical control circuit installation and adjustment, etc.. Through the practical training of this project, the trainees can meet the basic technical requirements for electricians at elementary level. The trainees have the technical theory and operation skills required by the junior workers of this type in *Standard for Workers' Technical Grade*, and can complete the routine work of this type independently using basic skills.

(II) Specific objectives

1. Professional competence

Have a certain learning ability and an ability of analysis, reasoning and judgment; have a certain expressive ability, planning ability and spatial sense; be able to distinguish various colors, have high flexibility of fingers and arms, movement coordination, no acrophobia. Have a knowledge of the safe use of electricity, regulations for low-voltage electrical devices, basic operation knowledge of fitters, basic mechanical knowledge, and metal technology knowledge.

2. Methodological competence

Have the ability to read electrical drawings and install devices according to electrical drawings to achieve simple electrical functions. Have a certain ability to identify and troubleshoot faults.

3. Social competence

Be able to be engaged in basic electrician related work, with the ability to analyze and solve complex problems. Team work and reasoning and judging ability are also required.

V. Training Contents and Requirements

The overall training contents include four training modules of 12 specific training tasks; each module has a number of exercises for extension. According to the specific needs and skills level of each college, modules and tasks can be selected and combined, and training programs and training duration suitable for the needs of the college can be formulated. In the course of training, trainees are required to take part in learning and training according to the training program, and complete the examination. Only through the examination can they obtain the training certificate.

VI. Suggestions on implementation

1. Suggestions on organizing the training

According to the needs of overseas technical training, local government or relevant departments will recruit students, and Chinese and Cambodian teachers of Lancang-Mekong project team will prepare lessons according to specific training requirements.

2. Suggestions on selecting textbooks

The special training textbook "Lancang-Mekong Vocational Education and Training Center Project Group" are recommend.

3. Suggestions on training methods

Integrated training and learning of theory and practice is carried out based on the systematization of work process and task introduction method.

4. Suggestions on development and utilization of training resources

Make full use of the training equipment of Lancang-Mekong Vocational Education and Training Center to carry out targeted technical skills training, and adopt project-specific training materials. At the same time, accumulate experience in the training process to improve or supplement the design of training modules and tasks.

5. Suggestions on configuration of experimental training

In Lancang-Mekong Vocational Education Training Center, Power Electronics and Electrical Drive Technology Training Room, Electrical Control Technology Training Room, Electrical and Electronic Technology Training Room, PLC and Sensor Application Technology Training Room are used for primary electrician training. Relevant equipment is prepared according to the training instruction materials.

VII. Assessment and Evaluation

1. Outline of assessment and evaluation

The training assessment should be carried out according to characteristics of the trainees and focus on the basic operational skills. The traditional assessment methods should be reformed and a scientific, flexible and usable project evaluation system should be established. The knowledge and technical ability and professional ability are comprehensively evaluated by the process assessment method of observation, operation, record, discussion, group evaluation and skill test.

2. Design of assessment and evaluation

(1) Attendance accounts for 30% of the total assessment result and the final assessment result accounts for 70%.

(2) The final assessment is based on the combination of theory and practice. Close-book examination is used for theoretical knowledge, accounting for 50% of the final assessment result; task-based training is used for practical operation skills, accounting for 50% of the final assessment result.

(3) The total score = 30% of the attendance + 70% of the final assessment score.

Appendix 4 Question Bank of Electronics (Elementary Level) (with answers)

Test One

I. Multiple-choice (Fill in the brackets with the letter before the correct answer. 1 point for each question, and 20 points in total)

1. The voltage of low voltage power supply is ( ).

A. 220 V B. 380 V

C. 220 V, 380 V D. 220 V, 380 V, 10 kV

2. The requirements for the quality of power supply workers in the basic code of conduct are: ( ).

A. study diligently, and be proficient in business

B. love electricity industry and be loyal to duty

C. obey the rules and discipline and be honest and self-disciplined

D. obey the leadership and obey orders

3. The number of electricity inspectors when conducting on-site inspections by electricity inspectors is( ).

A. at least one person B. no less than two people

C. no less than three people D. no matter how many people

4. The housing of electrical equipment is grounded, belonging to ( )

A. protection grounding Type B. lightning protection grounding type

C. working grounding type D. safety type

5. The school uses electricity for teaching and installs a self-contained transformer 315 kilovolt-ampere. The school's electricity fees should be charged according to ( ).

A. big industrial electricity price B. non-residential lighting price

C. non-industrial electricity price D. commercial electricity price

6. When a power supply bureau plans to sell 1.2 billion kilowatt-hours of electricity in 2002 and actually completes 1.5 billion kilowatt-hours of electricity, the completion rate of its electricity sales plan target is ( ).

A. 80% B. 125% C. 105% D. 100%

7. The switch of a multimeter is ( ).

A. a switch of various measuring types and ranges

B. a switch for current to go through the multimeter

C. a switch for connecting the measured object

D. a megohm meter switch

8. The oil of power transformer plays an ( ) role.

A. insulation and arc extinguishing

B. insulation and antirust

C. insulation, heat dissipation and arc extinguishing

D. insulation and discharge

9. When wearing insulating gloves for operation, the cuffs of the coat should be ( ).

A. put into the insulating gloves B. rolled up

C. put outside the gloves D. take off the coat

10. When an outdoor line switch is out of power for overhaul and traffic can pass through the bypass nearby, a sign ( ) should be suspended on the handle of the switch operator.

A. work here B. no closing

C. no climbing, high voltage danger D. watch out for falling objects

11. In the power supply, the direction of the electromotive force and the direction of the voltage are ( ).

A. the same B. the opposite

C. different by 90° D. arbitrary

12. For general metal materials, the resistance of conductors ( ) with the increase in temperature.

A. increases B. decreases

C. keeps constant D. is uncertain

13. The total capacitance of two capacitors of 10 μF in parallel is ( ) μF.

A. 10 B. 5 C. 20 D. 8

14. Connect the machine lighting bulb of a rated voltage of 24 V with a DC power supply of 24 V, ( ).

A. the bulb will be burned out B. the brightness is reduced

C. the bulb will be of normal brightness D. the bulb will be off

15. To measure the insulation resistance of electrical equipment, you should choose ( ).

A. bridge B. megohm meter

C. multimeter D. transformer

16. The frequency of AC in China's power grid is ( ) Hz.

A. 60 B. 50 C. 80 D. 100

17. The three elements of sinusoidal AC are ( ).

A. maximum, frequency, and initial phase angle

B. average, frequency, and initial phase angle

C. frequency, period, and maximum

D. maximum, RMS, and angular frequency

18. Transformers can be used to change ( ).

A. DC voltage and current B. AC voltage and current

C. frequency D. power

19. If the power wiring connecting any two-phase winding is exchanged, the motor will( ).

A. stop rotating B. turn invariant

C. reverse D. first reverse and then forward

20. The motor of a capacity below ( ) kW can be started directly.

A. 10 B. 7.5 C. 15 D. 75

II. Fill in the blanks (2 points for each blank, and 20 points in total)

1. Insulating materials in common electrical materials can be classified into _______grades according to the limit temperature.

2. The tool used to measure insulation resistance is _______. The insulation resistance of an AC motor should be greater than ________megaohms.

3. The main components of an electric fan are__________________.

4. The wiring principles of three-hole sockets are as follows____________________, ______________________,____________________________:

5. The common faults of a lighting circuit are mainly_____________, _______________ and_______________________.

III. True or false (Fill "T" or "F" in the brackets. 1 point for each question, and 20 points in total)

1. The direction of the electromotive force is from the positive pole of the power supply to the negative pole.( )

2. Contactless switch is a kind of position switch.( )

3. Circuit has three working states: path, short circuit and open circuit.( )

4. The direction of negative charge movement is defined as the direction of current.( )

5. When the reference point is changed, the potential of each point in the circuit also changes.( )

6. The magnitude and direction of steady DC current do not change with time.( )

7. When capacitors are connected in parallel, the equivalent capacitance is equal to the sum of the capacitances.( )

8. When using an ammeter, it is necessary to incorporate the ammeter into the circuit under test.( )

9. The three working states of a transistor are cut-off zone, saturation zone and amplification zone.( )

10. According to the internal structure of the transistor, it can be divided into NPN type and PNP type.( )

11. The output voltage of single-phase bridge rectifier circuit is equal to that of single-phase half-wave rectifier circuit.( )

12. The diode operates in the reverse breakdown zone.( )

13. For a DC electromagnet, if the armature is stuck in the process of suction, the coil will

be burned out.( )

14. The ballast of a fluorescent lamp (a coil with iron core) works on the principle of self-inductance.( )

15. The higher the frequency, the smaller the inductance.( )

16. The number of turns on the secondary side of a step-down transformer is less than that on the primary side.( )

17. The insulation level of a three-phase Y series asynchronous motor is B.( )

18. The stepless speed regulation can be achieved by changing the speed regulation method of asynchronous motor pole pairs.( )

19. Frequency converter can be used to change the speed of an AC asynchronous motor.( )

20. Reverse braking of asynchronous motors is suitable for large motors with frequent stop braking.( )

IV. Answer the questions (7 points for each question, 28 points in total)

1. Briefly describe the meaning of each item in the nameplate of the motor (in which the type of the motor is specified).

<table>
<tr><td colspan="3">Type: Y112M-4</td></tr>
<tr><td>4.0 kW</td><td colspan="2">8.8 A</td></tr>
<tr><td>380 V</td><td>1440 r/min</td><td>50 Hz</td></tr>
<tr><td colspan="3">Connection method △</td></tr>
</table>

2. Under what circumstances can electrical equipment cause space explosion?

3. What are the benefits of fast action of relay protection devices?

4. What should we pay attention to when we cut off the power supply before the power is cut off at the fire site?

V. Essay-type question (12 points)

Please illustrate the use of a multimeter and its precautions from the point of view of an electrician.

Keys to Test 1

I. Multiple-choice

1. C	2. B	3. B	4. A	5. C	6. B	7. A	8. C	9. A
10. B	11. B	12. A	13. C	14. C	15. B	16. B	17. A	18. B
19. C	20. B							

II.Fill in the blanks

1. 7

2. megaohms meter, 0.5

3. AC motor

4. left zero, right fire, upper grounding

5. circuit break, short circuit, leakage

III. True or False

1. F	2. T	3. T	4. F	5. T	6. T	7. T	8. F	9. T
10. T	11. F	12. F	13. F	14. T	15. F	16. T	17. T	18. F
19. T	20. F							

IV. Answer the questions

1. Y: motor series No.

112: height from motor base to axis.

M: base classification.

4: pole number.

4.0 kW: rated power.

8.8 A: rated current.

380 V: rated voltage.

1440 r/min: rated speed.

50 Hz: frequency.

△: connection method of three-phase stator winding and AC power supply.

2. Answer: Short circuit of equipment, lightning strike and insulation damage may cause space explosion.

3. Answer:

(1) Quick action, i.e. quick removal of faults, can reduce the working time of users in reducing voltage and accelerate the process of resuming normal operation;

(2) Quick removal of faults can reduce the damage degree of electrical equipment caused by faults;

(3) Quick removal of faults can prevent the expansion of faults.

4. Answer:

(1) After the fire, due to dampness or smoke, the insulation capacity of the equipment is reduced. Therefore, appropriate insulation tools should be used to pull the brake, and the electric shock should be prevented when the power is cut off.

(2) The place where the power supply is cut off should be appropriate to prevent the influence of the power supply cut off on the rescue work.

(3) When cutting off wires, different phase wires should be cut at different parts in order to avoid short circuit. When cutting air wires, the cutting position should be selected on the support in the direction of power supply, in order to prevent short circuit or electric shock accidents caused by falling down after cutting off wires.

(4) If there is a load on the line, the load should be removed first, and then the power supply on the fire extinguishing site should be cut off. When opening the switch of the brake knife to cut off the power supply, the tips are operated by using insulating rods or wearing insulating gloves to prevent electric shock.

V. Essay-type question

Key points involved are as follows:

(1) Setting at zero before use. In order to reduce errors, mechanical zero adjustment should be carried out before the multimeter is used. Before measuring resistance, ohmic zero adjustment should be made.

(2) Correct wiring. Before measurement, the multimeter red pen is connected with the "+" polar hole, and the black pen is connected with the "–" polar hole. When measuring DC current, the multimeter should be connected in series in the circuit. When measuring voltage, the multimeter should be connected in parallel across both ends of the circuit under test.

(3) Correct selection of measuring gears. When measuring voltage, current or resistance, the switch of the multimeter must use the corresponding measuring gear and make a good choice of measuring range. When the measuring range cannot be determined, the switch should correspond to the maximum range.

(4) Correct reading results. When measuring, we should read on the corresponding scale, and pay attention to the coordination of scale reading and measuring range.

(5) Pay attention to operation safety. In the process of high voltage measurement, we must pay attention to personal safety and instrument safety, and strictly prohibit live switching range switch. In addition, after use, the switch is placed in neutral or the highest AC voltage. If it is not used for a long time, the battery should be removed.

Test Two

1. Multiple-choice (Fill in the brackets with the letter before the correct answer, 2 points for each question and 20points in total)

1. In the power supply, the direction of the electromotive force and the direction of the voltage are (　　).

A. the same　　B. the opposite

C. different by 90°　　D. arbitrary

2. For general metal materials, when the temperature rises, the resistance of the conductor (　　).

A. increases　　B. decreases

C. remains unchanged　D. is uncertain

3. The total capacitance of two capacitors of 10 μFin parallel is (　　) μF.

A. 10　　B. 5　　C. 20　　D. 8

4. Connect the machine lighting bulb of a rated voltage of 24 V with a DC power supply of 24V, (　　).

A. the bulb will be burned out　　B. the brightness is reduced

C. the bulb will be of normal brightness　　D. the bulb will be off

5. To measure the insulation resistance of electrical equipment, you should choose (　　).

A. bridge　　B.megohms meter

C. multimeter D. transformer

6. The frequency of AC in China's power grid is ( ) Hz.

A. 60 B. 50 C. 80 D. 100

7. The three elements of sinusoidal AC are ( ).

A. maximum, frequency, and initial phase angle

B. average, frequency, and initial phase angle

C. frequency, period, and maximum

D. maximum, RMS, and angular frequency

8. Transformers can be used to change ( ).

A. DC voltage and current B. AC voltage and current

C. frequency D. power

9. If the power wiring connecting any two-phase winding is exchanged, the motor will( ).

A. stop rotating B. turn invariant

C. reverse D. first reverse and then forward

10. The motor of a capacity below ( ) kW can be started directly.

A. 10 B. 7.5 C. 15 D. 75

II. Fill in the blanks (2 points for each blank, and 20 points in total)

1. The transistor has __________PN junctions.

2. ________of mobile electrical equipment and hand-held electric tools (except Class III) must be installed.

3. ________is the shortened form of alternating current.

4. The motor often uses __________ as the overload protection.

5. In the low-voltage power supply system of China's power grid, the line voltage is ________.

6. In machine tool electric equipment, the red mushroom head button is used for_________.

7. The process of converting alternating current into direct current is called__________.

8. The total resistance of two resistors of 10 ohms in parallel is _______ohm.

9. Silicon steel sheets commonly used in iron cores of motors, transformers and relays belong to _________materials.

10. The rated current of the secondary side of the current transformer is _________ ampere.

III. True or False (Fill "T" or "F" in the brackets. 1 point for each question, and 20 points in total)

1. The direction of the electromotive force is from the positive pole of the power supply to the negative pole.( )

2. Contactless switch is a kind of position switch.( )

3. Circuit has three working states: path, short circuit and open circuit.( )

4. The direction of negative charge movement is defined as the direction of current.( )

5. When the reference point is changed, the potential of each point in the circuit also

changes.( )

6. The magnitude and direction of steady DC current do not change with time.( )

7. When capacitors are connected in parallel, the equivalent capacitance is equal to the sum of the capacitances.( )

8. When using an ammeter, it is necessary to incorporate the ammeter into the circuit under test.( )

9. The three working states of a transistor are cut-off zone, saturation zone and amplification zone.( )

10. According to the internal structure of a transistor, it can be divided into NPN type and PNP type.( )

11. The output voltage of single-phase bridge rectifier circuit is equal to that of single-phase half-wave rectifier circuit.( )

12. The diode operates in the reverse breakdown zone.( )

13. For a DC electromagnet, if the armature is stuck in the process of suction, the coil will be burned out.( )

14. The ballast of a fluorescent lamp (a coil with iron core) works on the principle of self-inductance.( )

15. The higher the frequency, the smaller the inductance.( )

16. The number of turns on the secondary side of a step-down transformer is less than that on the primary side.( )

17. The insulation level of a three-phase Y series asynchronous motor is B.( )

18. The stepless speed regulation can be achieved by changing the speed regulation method of asynchronous motor pole pairs.( )

19. Frequency converter can be used to change the speed of an AC asynchronous motor.( )

20. Reverse braking of asynchronous motors is suitable for large motors with frequent stop braking.( )

IV. Answer the questions(40 points in total)

1. Briefly describe the meaning of each item in the nameplate of the motor (in which the type of the motor is specified).

Type Y112M-4		
4.0 kW	8.8 A	
380 V	1440 r/min	50 Hz
Connection method △		

2. How many kinds of safe lighting voltage are there in common working area of machine tools under general environment? Name three kinds of asynchronous motor speed regulation.

3. Please list five types of relays you know.

Key to Test 2

I. Multiple-choice

1. A	2. B	3. C	4. C	5. B	6. B	7. A
8. B	9. C	10. B				

II. Fill in the blanks

1. 2
2. Leakage protector
3. AC
4. thermal relay
5. AC 380 V
6. emergency stop
7. rectification
8. 5
9. soft magnetism
10. 5

III. True or False

1. F	2. T	3. T	4. F	5. T	6. T	7. T	8. F	9. T
10. T	11. F	12. F	13. F	14. T	15. F	16. T	17. T	18. F
19. T	20. F							

IV. Answer the questions

1. Y: motor series No.

112: height from motor base to axis.

M: base classification.

4: pole number .

4.0 kW: rated power.

8.8 A: rated current.

380 V: rated voltage.

1440 r/min: rated speed.

50 Hz: frequency .

$\triangle$: connection method of three-phase stator winding and AC power supply.

2. 24 V, 36 V, variable pole speed regulation, variable slip speed regulation (variable resistance speed regulation, variable voltage speed regulation) and variable frequency speed regulation.

3. Time relay, intermediate relay, current relay (under, over), voltage relay (under, over, zero), speed relay, thermal relay, pressure relay, power relay, and solid state relay.

References

[1] 曾荣，王霞. 电工材料[M]. 北京：中国电力出版社，2007.
[2] 汤蕴缪. 电工学[M]. 5 版. 北京：机械工业出版社，2015.
[3] 孙方汉. 进网作业电工通用培训教材[M]. 2 版. 北京：中国电力出版社，2013.
[4] 苗玲玉，孙秀廷. 电气控制技术[M]. 2 版. 北京：机械工业出版社，2014.
[5] 徐旻. 电子技术及技能训练[M]. 2 版. 北京：电子工业出版社，2013.
[6] 张玉龙. 维修电工(基础知识)[M]. 2 版. 北京：中国劳动社会保障出版社，2012.
[7] 陈永强，魏金成，吴昌东. 模拟电子技术[M]. 北京：人民邮电出版社，2013.
[8] 康华光. 电子技术基础模拟部分[M]. 5 版. 北京：高等教育出版社，2008.
[9] 周忠. 数字电子技术[M]. 北京：人民邮电出版社，2012.
[10] 阎石. 数字电子技术基础[M]. 5 版. 北京：高等教育出版社，2006.
[11] THOMASL. Floyd. 数字电子技术[M]. 10 版. 北京：电子工业出版社，2014.